Praise for THE LONG VIEW

'The deep future is another country and its recent rediscovery by the philosophers Derek Parfit and William MacAskill, among others, has set off a movement that is gathering strength and recognition. Richard Fisher brilliantly summarises and warmly extends this transformed and transformative thinking about time. He is a captivating guide as he shows how our present is a minuscule blip on humanity's temporal map; how deep-time thinking not only lends even greater force to action on climate change, it enlarges generally our moral sense and our aesthetic possibilities. It has, he argues, implications too for our wellbeing, for there is profound satisfaction in conferring benefits on as-yet unborn people we will never meet. Fisher has a rich historical sense and *The Long View* is simply crammed with interesting ideas. This is a beautifully turned, calmly persuasive but urgent book.'

Ian McEwan, novelist and screenwriter

'A wise, humane book laced with curiosity and hope. It will open your mind and horizons – and leave you giddy at the prospect of all that we may yet become.'

Tom Chatfield, author of *How to Think*

'Urgent and profound. Richard Fisher's *The Long View* shows how thinking differently about time can change the world. The future begins with how we imagine it. This is essential reading for anyone who wishes to be a good ancestor.'

David Farrier, author of *Footprints*

'Moving beyond the abjection that dominates today's headlines about the future, Richard Fisher charts a hope-fuelled path toward the twenty-third century . . . An astounding read.'
Vincent Ialenti, author of *Deep Time Reckoning*

'Some of our greatest powers derive from our ability to think about the future. Fisher takes us on a fascinating tour of long-mindedness, and why it matters.'
Thomas Suddendorf, co-author of *The Invention of Tomorrow*

'Philosophical, insightful and entertaining. In a world of short-sightedness, *The Long View* is a helpful guide to understand and connect us to the future. In the light of the climate emergency, long-term thinking is more urgent than ever.'
Andri Snær Magnason, author of *On Time and Water*

Richard Fisher

The Long View

Why We Need to Transform How the World Sees Time

WILDFIRE

First published in 2023 by
WILDFIRE
an imprint of HEADLINE PUBLISHING GROUP

1

Cataloguing in Publication Data is available from the British Library

Hardback ISBN 978 14722 8521 8
Trade paperback ISBN 978 14722 8522 5

Typeset in Dante MT by CC Book Production

Printed and bound in Great Britain by Clays Ltd, Elcograf S.p.A.

HEADLINE PUBLISHING GROUP
An Hachette UK Company
Carmelite House
50 Victoria Embankment
London EC4Y 0DZ

www.headline.co.uk
www.hachette.co.uk

For Clive Fisher

Contents

PART III

THE LONG VIEW:
EXPANDING OUR PERSPECTIVE OF TIME

Introduction:
A Longer View

'If we are to better the future, we must disturb the present.'
CATHERINE BOOTH

One February night in a London hospital, my perception of time shrunk to the span of a moment.

Around twenty-four hours after my wife went into labour, we were rushed into emergency surgery. Our baby had acquired an infection. As the doctors operated, it felt like nothing else in the world existed. During crisis, you can only focus on the present.

As my wife recovered and our new daughter, Grace, completed a course of antibiotics, the two of them stayed in the maternity ward for a week. It was a period of no before or after, no yesterday or tomorrow, only the immediacy of new parenthood and the stress of a little girl who was unwell.

Eventually, we returned back home. As the days passed, our new daughter's hands reached out of the Moses basket, her grey-blue eyes sharpened, and a few weeks later, she gave us the reward of a smile. My awareness of time returned – and I began to think about the life that lay ahead for her. The child she could become, the adult she'd grow to be.

What I had never considered until then was that Grace could live to see the twenty-second century. She'll be eighty-six, which is not improbable with advances in healthcare and longevity. I imagined her on New Year's Eve, welcoming 2100 with her own family as the fireworks explode in the sky, loved ones embrace, and 'Auld Lang Syne' is sung.

It's easy to forget that there are millions of citizens of the twenty-second century already among us. In the year of my daughter's birth, 140 million other people were born worldwide, more than twice the population of France. Looking ahead, almost eleven billion more children will begin their lives before 2100. And if they are lucky, some of them – including Grace's grandchildren – will live through to the *twenty-third* century. Through our family ties, the deep future is far closer to the present day than it first appears.

Later, though, I could not help but feel troubled by the daydream about my daughter reaching the final New Year's Eve of the century. Throughout my career as a journalist, I have encountered this date – 2100 – within various news stories, reports or forecasts, but they are rarely celebratory.

Rising Seas and Wild Weather Threaten to Drown Cities by 2100
Humanity has 1 in 6 Chance of Extinction by End of the Century
Nearly Half the Planet's Species Could be Wiped Out in 2100
Automation Will Leave Humans Jobless by Twenty-Second Century

The year 2100 is so often depicted as a milestone within a worsening world, a marker for society's decline, or sometimes even a barrier we may not pass. Such stories of future climate turmoil, biodiversity collapse and technology disruption often feel so far away in time, but we are connected to that turning point by only one or two generations.

In the 1700s, the political thinker Edmund Burke wrote that society is a contract: 'a partnership not only between those who are living, but between those who are living, those who are dead, and those who are to be born.'[1] Unfortunately, this partnership between the generations is breaking. If our descendants were to diagnose our generation's most harmful habit, they would observe a dangerous new form of short-termism – particularly in the West. In the early twenty-first century, the 'now' commands all attention, and any sense of past and future is filtered only through the events of the current moment. The world is saturated by information, and standards of living have never been higher – yet so often it's a struggle to shift attention beyond the next news cycle, political term or business quarter. If time can be sliced, it feels that it is getting ever finer.

But what makes this strain of short-termism so pernicious is its invisibility. It goes far beyond the conscious, deliberate prioritisation of near-term goals. This is short-termism without awareness: a blinkered perspective that permeates capitalism, politics, media and popular culture. Not only are the lessons of history being ignored, but also how our acts are rippling into tomorrow.

Living in the age we do, we have never before had such leverage to shape the trajectory of the future, with so little collective recognition of that fact. We now have technology powerful enough to cause a catastrophic setback to civilisation, through the force of the atom, biological warfare or digital creations run amok. And for the first time, the human species has the ability to irreversibly degrade the planet's biosphere and climate.

Every year that the blinkered perspective dominates, the risks escalate: climate change, infectious disease, biodiversity collapse, antibiotic resistance, artificial intelligence, or nuclear war. Meanwhile, inequality grows, healthcare costs rise and infrastructure rots, and our children become ever-more burdened with an inheritance of malignant heirlooms that they are obliged to accept, from plastic fibres to atomic waste.

We live amid a precarious time of crises, during which it feels more difficult than ever to escape the priorities of the present. However, I do not believe that this should be cause for helplessness or resignation. Humanity is capable of evolving to a better understanding of our roles and responsibilities within the long term. History shows that people's perceptions of time have transformed before, so there is no reason that they cannot transform again.

*   *   *

The Long View has a simple aim: to understand why the world became so blinkered to the long arc of time, and how to change that perspective. It tells the story of what led twenty-first-century civilisation into a period of short-termism, why people think about the past, present and future in the way they do, and how to become more 'long-minded'. We tend to look at the world through various well-established lenses – political, social, financial – but we need a perspective that is *temporal* too.

The book offers an intellectual framework that will enable you to see the perspective-shortening forces hidden in plain sight within your surroundings, and the tools and approaches to overcome them. A litany of invisible short-termist incentives and deterrents have come to influence business strategy, political policies, media coverage and the choices made by individuals. I call these *temporal stresses*. But crucially, none of these pressures is necessary for a functioning society, and all can be avoided if we can identify what they are.

The lens on the world offered by this book also allows you to spot the *temporal habits* within your own psychology, and that of others. We are only at the beginning of understanding how our brains process time, but what we do know is that people's decisions are influenced by subtle cognitive traps that can shorten their perspectives. These biases often operate beneath conscious perception – but none is impossible to tackle if brought into the open. While it may sometimes feel difficult to escape the present, it is not beyond our mental faculties to do so.

After all, many around the world are already unlocking the long view. These long-minded individuals, organisations and cultures are scattered across the realms of capitalism, governance, art, history, philosophy and technology – but what unites them is the discovery that a long-term lens brings myriad positive benefits to themselves and others. Their stories show why embracing a deeper perspective of time can be transformative.

In a period when societies are grappling with climate change, pandemics, inequality and political upheaval, the long view is needed more than ever. If we continue on the path we are on, staring only at what's immediately in front of us and nothing else, we will inevitably stumble towards disaster. A longer vantage point, however, is available. From here, history emerges as a landscape of untapped wisdom and experience for tackling the urgent problems of today, and tomorrow reveals itself as far more open than it seems. It is a perspective that allows us to see just how far we have come and what we have learned along the way – but also shows us paths to a potentially better world. The blinkered age we live within hides the threats ahead, but it obscures *possibilities* too.

Some might say we are cursed to know that our lives are just short links in a chain from ancient ancestors to tomorrow's descendants: a mere flash in the history of life, our species and the Earth. How humbling it is to know that so many of our experiences and achievements will be forgotten only a few generations beyond our grandchildren. Yet it is also a blessing: part of what makes us human is contemplating the possibilities and potential of our trajectory through time, while handing down all we have learned and know about the world to the people that come next. For me, that is the great benefit of embracing the long view: that it reveals how the closest relationships in my life extend from the past into the future. And through that comes a sense of potential – and hope.

When my daughter and the rest of the world's descendants reach the twenty-second century, I would like them to know that ours was the generation that changed its ways and became more long-minded. That we learned from our mistakes, that we decided to reorient our priorities, and that we had the courage and wisdom to transcend our blinkered age and embrace the long view.

The story of the long view

The Long View is inspired by my research during a fellowship at Massachusetts Institute of Technology in the US. There, I aimed to understand the roots of the world's short-termist predicament, the history and psychology of how we think about time and the future, and the philosophy and ethics of the long term. But its origins go back further.

Throughout my life, I have been fascinated by humanity's place within deeper time. As a boy, I collected rocks and fossils, and at university I studied geology. The long term has also been a theme running throughout my career in journalism. More than a decade ago, I commissioned a special issue for *New Scientist* called 'The Deep Future'.[2] In it, we tackled a series of questions about humanity's next 100,000 years: Will we still be here? (Probably.) How will we evolve? (It depends on technology.) Where will we live? (Climate change will redraw cities, borders, nations.) Will there be any nature left? (Yes, but expect super-evolved pigeons and rats.) How will language evolve? (By 1,000 years, you won't understand it.) And what will future archaeologists know about us? (Plenty, but they'll be most interested in our garbage.)

More recently as an editor at the BBC, I led a digital series called 'Deep Civilisation', where I asked academics and writers to project their ideas into longer-term time, exploring the long view of democracy, religion, societal collapse, human intelligence, architecture, ethics, nature, technology and more.[3] I also produced a live discussion about long-term thinking at the Hay Festival, which was broadcast on BBC News. To date, the Deep Civilisation series has reached more than 6.5 million people.

Along the way, I have also encountered a growing number of long-minded individuals, organisations, academics, foundations and politicians who are converging on the belief that it's possible to unlock a longer-term perspective. Over the past few years, I've aimed to immerse myself within this new burst of activity, and I write about what I'm learning in a publication and newsletter called *The Long-termist's Field Guide*.

So, what will follow in the pages ahead is a story that starts in the past, diagnoses the present and offers a new route into the future. By weaving together case studies, academic research, history, science and philosophy, I will answer three central questions, across three parts: How did we get here? Why do we think the way we do? And where should we go next?

Part I

TIME-BLINKERED:

THE ROOTS AND CAUSES OF SHORT-TERMISM

In Part I, we will explore how modern societies became blinkered to the long view, particularly in the West. If we can understand why this happened, the route out starts to become clearer. Finding those answers begins with a 'brief history of long time': a journey across the past few thousand years, taking in ancient timekeeping, the Roman empire, the Middle Ages, the utopian ideals of the 1700s all the way through to the tumultuous twentieth century. How did societies in the past think about longer-term time, and how were those perspectives influenced by culture, religion and scientific discovery?

Next, we will turn to some of the most dominant pressures that nudge present-day societies into the short term – and how these *temporal stresses* have grown in influence. We'll look at the worlds of business and economics, and then governance and media, and explore how the pressures within these realms of modern life have supercharged short-termist tendencies and exacerbated risks far beyond what our ancestors dealt with. Along the way, we'll also meet the individuals and organisations from around the world who have resisted these stresses: the CEOs who have defied shareholders, the companies that have persisted for 1,000 years, the politicians campaigning for future generations, and more.

Part II

A TEMPORAL STATE OF MIND:

UNLOCKING OUR AWARENESS OF TIME

In Part II, our goal will be to gain a deeper understanding of how time is perceived by the human brain. If the first part of our story is about the external pressures fostering blinkered behaviour, in these chapters we will look inward, to understand the processes – and contradictions – within our own minds. These are the *temporal habits*.

First, we'll trace how we developed the mental skill of transporting our conscious self across time. How and why did our unique form of time perception evolve and develop? The aim here is to understand how the brain constructs the abstract idea of past and future, and how this skill compares with the abilities of other animals. Next, we'll explore the lessons of psychology, and the cognitive biases that can sway people's behaviour and decisions to prioritise the present moment. What are the psychological barriers to long-term thinking, and how do we overcome them? And finally, we'll look at why the language we use to describe the past, present and future matters. Each tongue expresses time differently, and it may well be that this is subtly influencing short- and long-term thinking around the world.

Part III
THE LONG VIEW:
EXPANDING OUR PERCEPTION OF TIME

In Part III, the goal is to derive insights from the people, projects, cultures and organisations that are showing the way to greater long-mindedness. While the long view may not be widespread within the time-blinkered societies that many of us live within, it's there if you look.

We'll begin by exploring the sublime emotions evoked by deep time and the benefits that come from embracing it, as well as identifying the surprising connections we share within our family trees over the long term. Then we'll look at alternative timeviews, beginning with the perspectives of religion, the significance of long-term rituals and lessons in intergenerational reciprocity from indigenous groups.

Next, we'll discover a long-minded perspective that has emerged more recently: an ethical approach called *longtermism*.[4] This is a call to reconsider our moral obligations to the future and involves dizzying calculations of the scale of tomorrow's generations. It's a timeview that suggests our species may just be at the beginning of a far longer trajectory.

We'll also dive into the long view of nature: how scientists have unlocked *temporal windows* to the past and future of the natural world, taking in the Big Bang, plate tectonics, the discovery of global warming and the dawn of the Anthropocene. And our quest for the long view would be incomplete without learning from artistic and symbolic endeavours: libraries that won't publish their contents until the year 2114, musical compositions running for 1,000 years and an artwork fashioned from trees that went unexpectedly wrong.

So, let's begin our path towards a deeper temporal perspective of the world. In the coming pages, we'll traverse the past, present and future of humanity and the planet, taking in the history of our species, a broader understanding of the mind, and the people and cultures that point to a long-minded future. With these insights, a richer – and more hopeful – lens on the world awaits. Welcome to the long view.

PART I

TIME-BLINKERED
THE ROOTS AND CAUSES OF SHORT-TERMISM

I

A Brief History of Long Time

'The way in which time and its horizons are conceived is generally connected with the way the society understands and justifies itself.'

ERNEST GELLNER[1]

If you could travel back thousands of years to ask your ancestor what they thought about time, what might they say? Would they have had a long view?

Physiologically, your ancestor would have had the same brain as you, capable of remembering yesterday and thinking about tomorrow. However, we can make a reasonable guess that their perspective of longer-term time would have been different, shaped by their knowledge, culture, beliefs and assumptions about the world.

You and I can picture ourselves on a linear timeline that stretches from the Big Bang into the deep future, populated by moments of flourishing and catastrophe, such as the origin of life, the rise of *Homo sapiens* and the eventual death of the Sun. Science has revealed that our species is a relatively late entrant in this story: Earth was here long before us, and the Universe will outlast us when we are gone. Meanwhile, technology has given us clocks, calendars and stratigraphic charts that display where we sit within time's long arc: it is 4 p.m. on a Tuesday in the second millennium, somewhere around the dawn of the Anthropocene.

Our ancestors would not have shared our particular perspective, but this does not mean they did not have their own version of the long view. After all, history features many acts of grand forward planning: Stonehenge, the Pyramids and cathedrals. And for thousands of years, religious followers have believed in an eternal afterlife, or immense epochs that continue repeating forever.[2] Through their eyes, time would seem to stretch to infinity.

In the 1960s, the philosopher Ernest Gellner pointed out that each society develops a different perspective of time, which influences its decisions and trajectories. Across history, some have had an 'unchanging temporal horizon, like a train crossing a featureless landscape', he wrote, while others 'live in the anticipation of the ending of time and conceive of its value as a preparation for that termination'.[3] And some have, occasionally, managed to embrace a longer view of time that transcended their own circumstances.

So, what might the temporal perspectives of our ancestors tell us about our own? We have become the architects of grand metropolises and masters of science and technology, a trajectory of progress that might appear to be interwoven with an ever-expanding sense of time. But that would be an oversimplification. If the story were so neat, our horizons in the twenty-first century should reach further than ever before, but in practice that's not the case.

It may be true that, compared with our ancestors, we have a clearer sense of our position in deep time. But that knowledge rarely figures in everyday culture. While we may have the ability to access an array of histories and possible futures, they are often unwittingly viewed through the lens of the present, shaped by the concerns and priorities of the moment. Why are we so stuck in the 'now'?

To answer this question, we must plot how we got here in the first place. Only then can we begin to navigate a route forward. The story that follows charts how temporal perspectives have changed in the West over the past 2,000 years or so. It's not the only history – we'll discover alternative global views later in the book – but understanding it is important because the attitudes to time entwined with Western culture now affect the entire world.

So, let's begin, with a man who was irritated by a sundial.

The parasite and the sundial

In the second century BC, the Roman playwright Plautus wrote a comedy featuring a character called the 'hungry parasite'. In classical theatre, the parasite archetype was a hanger-on, or a house-guest taking advantage of his host's hospitality, getting by on wit and manipulation.[4] You might imagine him an obsequious but charming slob.

In this play, the parasite had a gripe about a piece of new-fangled technology that was interfering with his appetite:

> The gods confound the man who first found out
> How to distinguish hours. Confound him too,
> Who in this place set up a sundial,
> To cut and hack my days so wretchedly
> Into small pieces! When I was a boy,
> My belly was my sundial – one surer,
> Truer, and more exact than any of them.
> The dial told me when 'twas proper time
> To go to dinner, when I ought to eat:
> But now-a-days, why even when I have,
> I can't fall to unless the sun gives leave.
> The town's so full of these confounded dials.[5]

What these lines of Plautus' play captured was more than a parasite's hunger. They also showed that the Romans of this period were transitioning to a broader shared sense of time.

In a town full of 'confounded dials', the parasite complained, his meal-times were set, stopping him from eating when he pleased.[6] No longer could he rely on his own belly as a method of timekeeping.

Roman sundials weren't especially accurate: their hour ranged from around forty-five to seventy-five minutes, depending on whether it was winter or summer. But it was an important change. When societies adopted an independent sense of time that existed outside each human mind, then it allowed for a more precise shared chronology of past, present and future. It also allowed many other developments to happen – not just the setting of mealtimes. If each individual kept their own time, they would have struggled to collaborate; a shared one meant people could assemble to build, plan and organise.

The Romans were not the first timekeepers. The Ancient Egyptians had built obelisks, whose shadows indicated the rough time of day and season. They also deployed water clocks to measure duration. The early Chinese dynasties had similar inventions.

By this period, calendars had also long since emerged. Before the Romans, the Sumerians, who lived in modern-day Iraq, had split their year into thirty-day months and twelve-period days. The Egyptian calendar had three seasons of 120 days each. And Chinese astronomers had been the first to intercalate and reconcile the differences between the lunar and solar cycles. According to legend, China's Yellow Emperor invented a calendar around 4,500 years ago.

For the Romans, their original calendar is believed to have been based on the Moon. Their Julian calendar was introduced in 45 BC, becoming the predominant calendar in Europe for around 1,600 years.

So, at the turn of the first millennium, human beings in Europe and elsewhere had embraced basic timekeeping devices, and calendars that progressed in a linear fashion. If you could step into the shoes of an Ancient Roman, would they have had a long view?

The Romans had a language that could describe the events of yesterday and tomorrow and used the same spatial metaphors of the past being 'behind' and the future 'in front'. You could also see evidence of impressive forward planning: they built long-lasting roads and aqueducts that catered to their future needs. Yet on a societal scale, Roman perceptions of time may have been different to our own – particularly their view of the long-term future.

According to the historian Brent Shaw of Princeton University, the Roman future may have been less elaborate, less deep and more fragmented than in modern societies, and more dependent on personal connections and immediate concerns.[7] The Romans may have known intellectually that there was a world beyond the present, but he argues that their future was 'adumbrated' – only faintly drawn.

'One dominant concept was the idea that time was flowing toward them. That the future was a fixed thing,' Shaw told me. Tomorrow was not a complex space of multiple scenarios and branching possibilities, but if anything a single, predetermined track. And like many other ancient cultures around the world, the Romans clung to the concept of fate, personifying the concept as goddesses who shaped human destiny.

The Ancient Greeks may have shared a similar perspective. Even the great thinker Aristotle saw human civilisation as static, reasoning that everything that could exist, already existed. To him, 'all discoverable things have already been discovered; all thinkable things previously thought; all forms of government already assessed; all workable feats of engineering tried and tested', writes the historian Thomas Moynihan.[8] What this implies is that Aristotle and his contemporaries had little sense that there could be a future world with ideas, technologies and things that were not already part of their universe.

All this could explain why the Romans and Greeks turned to mysticism. Throughout history, ancient societies in Europe, China or Mesopotamia sought the wisdom of oracles, who promised answers from entrails, fire, dreams, bones, or even the cracks on roasted turtle shells.[9] And the Romans were no different. Before officials embarked on battles or elections, they'd ask a priesthood of augurs to study bird behaviour for guidance. (In Latin, the word *'auspices'* essentially means 'looking for fortunate signs in birds'.[10]) With the confidence of the historian relating fixed events in the past, these augurs would provide information about fixed events in the future.

Perhaps if the Romans and Greeks had continued to thrive, their perspective would have changed. But around the same period, another temporal lens on the world was emerging, via a religious belief that would lock in a foreshortened view of human time for more than 1,000 years.

Apocalypse, now

On the night of 18 July 64CE, a fire broke out in the Circus Maximus in Rome, close to the Palatine and Caelian hills, and burned for many days. Whipped up by the wind, it consumed buildings, shrines and temples, outstripping 'all defensive measures because of the speed of its deadly advance', wrote the Roman historian Tacitus.[11] Panic followed. 'Some chose death because they had lost all their property, even their daily livelihood; others did so from love of family members whom they had been unable to rescue.'

Faced with such a disaster, the emperor Nero needed someone to blame. Many suspected he had been involved – which helps to explain the apocryphal story of his fiddling while the city burned. A convenient scapegoat? The Christians.

Nero waged a terrible campaign on this fledgling religion: execution by crucifixions, burning alive and dogs. His cruelty was so great that he would be written in to the apocalyptic Revelation to John, an anti-imperial book of the New Testament, which called for God's intervention to overcome political oppressors. It features a seven-headed beast rising from the sea, which brings about the end of the world. Intriguingly, this monster was also given a number. 'Let him that hath understanding count the number of the beast: for it is the number of a man; and his number is six hundred threescore and six.' Centuries later, scholars would realise that this number – 666 – was not random. It was a figure calculated using the Hebrew letters for 'Nero Caesar', which can be converted to 50, 200, 6, 50, 100, 60 and 200, adding up to 666.[12]

To early Christians, Nero represented more than a political oppressor. In their belief system, he was also a key figure in their conception of humanity's future on Earth – specifically, that there wasn't one. As the philosopher Gellner wrote, some societies live their entire existence in 'the anticipation of the ending of time' – and here was a clear example. When Christians looked beyond the present, they did not see a long and distant future stretching ahead of them – but near-term rapture and chaos, followed by an altogether different form of timeless existence alongside God.

While the expectation of religious eternity after the apocalypse was certainly one type of long view, it was different from a long linear 'timeline' perspective. Heaven was arguably more of a perpetual, transcendental version of the present. Painting a picture of infinity in the mind is easy if it is all coloured with the same shade. Or to borrow the title of an Emily Dickinson poem: 'forever is composed of nows'.[13]

The fact that the apocalypse did not arrive as promised did little to affect people's beliefs. When Nero killed himself in 68CE, expectation of the end-times did not die with him. In fact, many doubted his death: some assumed it was faked, while others thought he would be reborn. A handful of imposters and mistakenly identified lookalikes cemented the view that he would return. Later, a new antagonist was found: it was the Roman emperor Domitian doing the devil's bidding.

While the roots, timing and locations were different, apocalyptic expectations emerged within other religions too, featuring common themes such as cosmic signs, warnings of social degeneration, saviours and villains, and periods of suffering. Eschatological ideas can also be found in Islam and Buddhism. The Qur'an features a day of judgement (Yam al-Dīn), resurrection and a fight against al-Dajjāl (the Antichrist). And while Zen Buddhism might reject the premise of end-times – there is no beginning or end, only now – a prophecy in the Tibetan Kalachakra Tantra talks of an apocalyptic battle between barbarians and Buddhists led by the final king of Shambhala.[14]

As the Christian faith grew, a sense of the looming end-times became ever-more dominant in Europe, and persisted for more than 1,000 years.

In the 1100s, the Italian theologian Joachim of Fiore, an important apocalyptic thinker of the medieval period, added a new twist. Reinterpreting the Revelation of John, he argued that the world had three eras: the Age of the Father (the Old Testament), the Age of the Son (Jesus's life and the New Testament era) and a coming Age of the Spirit. Joachim's calculations suggested that he was living on the cusp of the third age, and a time of great change.

In Joachim's eyes, global transformation and the final defeat of the Antichrist was right around the corner, potentially beginning around 1260 or 1290, approximately a century from the time he was writing. Again, here was a perception of a future for humanity that was foreshortened by a short-term prediction of upheaval. This wouldn't be the end of the world – what he labelled *'finis mundi'* – but crucially it promised an abrupt disruption of the status quo.

In some ways, doomsday thinking has had something of a resurgence in recent decades. Depending on the mood of the age, art and culture often draw on the apocalypse in secular storytelling: *The Day of the Triffids, War of the Worlds, The Walking Dead, Terminator, The Day the Earth Stood Still* and many more. Many of these stories feature millennia-old religious ideas: a day of catastrophe, a single figure that sees it coming, a sacrifice and a saviour.

The downside of this doom-laden view of the future is that it can breed a nihilistic form of short-termism. When the apocalypse is nigh, the temptation is to party like it's the end of the world, or to give up trying to stop it. In the context of climate change, the scientist Michael E. Mann has called this 'doomism': the dangerous belief that acting to reduce the threat of runaway climate change is pointless because it's already too late.[15] (This is an important idea that we'll return to in later chapters.)

A time of cycles and cathedral thinking

As the late Middle Ages arrived, religion and time became entangled in other intriguing ways. Around the thirteenth century, timekeeping technology made another leap forward, with the invention of the escapement, which allowed clock designers to build mechanisms with gears that 'ticked' one tooth at a time. While these more advanced clocks promised an ever-more precise view of shared time for societies, it would be centuries before they were widely owned in households. Instead, many of the first timepieces in Europe rang as bells in churches, calling to prayer. And the world's first mechanical clocks can be found in Salisbury and Wells cathedrals, dating to the late 1300s. Time, for many, was owned by the Church.

The construction of the buildings that held these clocks can also provide us with hints about how people viewed the future during this period. By the 1300s, communities were often building their cathedrals as multi-decadal projects, potentially lasting beyond the lifetimes of those that kickstarted the project.

Work on Wells Cathedral, for example, began around 1175 and continued until 1450. The bishop who is thought to have conceived the project, Reginald Fitz Jocelin, died in 1192. Taking up the baton, Jocelin of Wells continued building in the early thirteenth century. By 1239, he had managed to complete the main part of the church, but made his own ascent to heaven three years later. He would not see the additional extensions, heightening and rebuilding that continued for more than a century hence.

Such 'cathedral thinking' is often held up as an example of our ancestors' admirable long-term view. What could be more long-term than designing a structure you know you will not live to see completed? Yet a closer look reveals that it was not quite as straightforwardly long-sighted as some in the present-day might like to believe.

Standing inside a grand cathedral that took many lifetimes to build, it's tempting to assume it emerged from a single, far-sighted blueprint. No doubt each one began with a vision, but many multigenerational building projects evolved more organically and haphazardly than their final appearance might suggest.

The same can be said of earlier human constructions that took decades or centuries to finish. Assembling Stonehenge, for example, involved remarkable forward planning, and its builders would have had a long view of their own: the site's sacred usage as a cemetery suggests these ancient people clearly valued the links between generations. However, its 1,500-year construction took shape in distinct stages, beginning with ditches and wood, and evolving through various forms over time. The Great Wall of China, too, was completed section by section, by several unconnected dynasties, and by different officials in response to local needs.[16] Grandeur can foster an illusion of grand design: if aliens landed on Earth today, and looked at all the world's metalled roads, or the global railway network, they might make similar assumptions about a long-sighted plan, when the reality is a patchwork of uncoordinated engineering.

In the present day, we also only see the architecture that survived the centuries, so ought to be cautious about assuming that it was representative. In the case of medieval cathedrals and churches, many were plagued by shoddy workmanship, fast-decaying materials or short-termist disputes over funding. We know this because master masons called 'viewers' – a bit like building inspectors – were charged with preventing such practices, and their intervention would often lead to lawsuits, and demands for total reconstruction. In the thirteenth century, Meaux Cathedral in France was ripped down and rebuilt due to bad planning and poor work.[17]

Even with the best intentions, Christian architects often failed to prepare their structures for longer-term adaptations. At Salisbury Cathedral in the UK, a 5,900-tonne tower and spire – added half a century after the main structure was completed – was too heavy for the original building to bear, and caused the giant central supporting columns to sink.

Fears of church and cathedral collapse in the medieval period were so pronounced that in some British places of worship a prayer would be said during services: 'Deare Lord, support our roof this night, that it may in no wise fall upon us and styfle us. Amen.'[18]

Finally, it's worth remembering that if cathedral-builders did have a long view, it would have been rooted in a different societal context to our own. Perhaps it was easier for them to imagine baton-passing into the future because change happened more slowly. They would have expected their descendants to have essentially the same life as them, with the same needs and desires.[19]

'In medieval times, most human affairs had the form of endless repetition: sowing and harvesting, disease and health, war and peace, the rise and fall of kingdoms – there was little reason to believe in long-term change or even improvement in human affairs,' writes the historian Lucian Hölscher at the University of Bochum.[20] If people had a perspective of the future, 'it did not involve the expectation of anything new, or change the image of a distant period of time. The cathedral would still be standing at the end of the world, but the scenery surrounding it would look essentially the same.'

So, while cathedral-building might look like a radical development in the perception of time, people may still have been living in more of a perpetual, cyclical present. They could conceive of a human world continuing after their death, but it resembled the static 'now' they lived within.

The discovery of the long future as an entirely different land was yet to come.

Statistics and goblin treasure

By the seventeenth century, the Western sense of time had slowly begun to evolve within a few domains, particularly governance and commerce. For leaders, it was no longer possible to rely on past events and traditions as a guide to rule, because social, political and cultural change was now happening too swiftly. History was coming to be seen as quite different to the past and the future, and so cyclical time was gradually evolving to be something more linear.[21]

A number of governments became increasingly concerned with facts and figures to help shape future expenditure and taxation,[22] and political advisors began making the first attempts to predict the growth of future populations. In 1696, the British statistician Gregory King predicted that world population would reach 630 million by 1950 and 780 million in 2050.[23] (King might have been shocked to learn that it would actually be 2.5 billion in 1950, and is projected to be 9.7 billion in 2050.[24])

Meanwhile, the rise of stock markets and the trading of company shares led to what we now call the language of 'futures'. In 1688, the oldest known book about the stock exchange was published, called *Confusion of Confusions*. It contained prescient wisdom: 'He who wishes to become rich from this game must have both money and patience.' And surprisingly poetic advice: 'Profit in the share market is goblin treasure: at one moment, it is carbuncles, the next it is coal; one moment diamonds, and the next pebbles. Sometimes, they are the tears that Aurora leaves on the sweet morning's grass, at other times, they are just tears.'

By this period, maritime insurance to protect against future losses had also been introduced in European ports, and life insurance had been developed in Amsterdam, following research on the mathematics of probability. As well as evidence of forward planning, these financial inventions reflected the idea that multiple futures were possible and to be successful one needed to prepare for various scenarios to play out.

Still, for the average person, religious teachings about the history and future of humanity remained one of the strongest accounts of longer-term time. The prospect of the coming apocalypse had not gone away and continued to assure believers of their convictions during political and social turmoil, such as the events before and after the English Reformation.[25] Sir Isaac Newton believed that the interpretation of biblical prophecy in books like Revelations was 'no matter of indifference but a duty of the greatest moment'. By some accounts, Newton privately calculated that the world would end in 2060.[26]

Meanwhile, most people still thought the world was only a few thousand years old. In 1650, the bishop James Ussher had published a calculation of the Earth's age that he believed provided the final word, dating creation to 4004BC using a literal reading of the number of generations described in the Old Testament.

Those biblical perceptions of the past and future, however, were about to be transformed.

Disconforming unconformities

In the second half of the eighteenth century, the dominance and credibility of the Bible in shaping the West's sense of time began to crumble, as the natural sciences uncovered striking evidence that the Earth was much, much older than theologians thought. One of the biggest steps in changing this view was a seemingly blasphemous claim from the Scottish geologist James Hutton, a man who would transform how we see ourselves within Earth's chronology.

I first learned about Hutton when I was sixteen or so, studying geology at school. On a field trip to Scotland's Isle of Arran, our teacher, Mr Veevers – red-faced in a waterproof jacket – would lead us up mountains and along coasts to gaze at exposed rock. One day, we visited a formation called 'Hutton's Unconformity'. To a teenager, it didn't seem like anything special, but when we looked closer there were two types of rock of very different ages, separated by a line. What we were staring at, our teacher assured us, is among the best examples of deep time that you can see with your own eyes – and in 1788, Hutton was the first to notice it.

Hutton's early life was focused on medicine and chemistry, and he also was something of an entrepreneur, manufacturing a profitable form of salt, useful for dyeing, among other things. But later he retreated to a life of farming in Scotland. There, he became fond of studying the soil and rocks around him. One day on Arran, he noticed a puzzling arrangement of the rocks in the outcrops he visited, which he called an 'unconformity'. The following year, he would demonstrate the significance of his observations by leading a group to Siccar Point, on the opposite side of Scotland. There, he pointed out another example of an unconformity that was clear to see: layers of grey rock arranged like vertical cards, sharply overlain by gently sloping red sandstone.

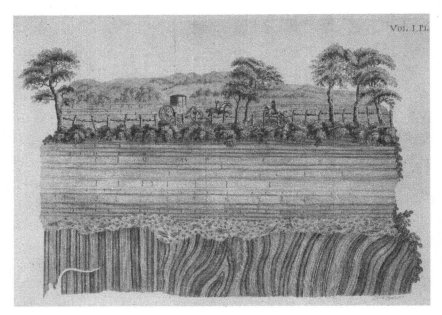

Hutton's Unconformity at Inchbonny, Jedburgh,
drawn by his friend John Clerk of Eldin.

Why did this matter? The only way they could have formed in this arrangement, Hutton showed, was via a process that lasted tens of millions of years. The layers below started as horizontal sedimentary rocks in an ancient ocean, before being buried and folded into their steep arrangement deep within the Earth's crust. Later, Earth's tectonic movements would have thrust them to the surface again to be eroded over many aeons. Then, one day, a new rock would begin to be laid down grain-by-grain – like a dry, red overcoat – eventually forming gently sloping beds of sandstone.

There simply wasn't enough time in the Church's account of a 6,000-year-old Earth for these rocks to have formed in this arrangement. 'The mind seemed to grow giddy by looking so far back into the abyss of time,' wrote John Playfair, one of Hutton's collaborators, who went on the trip to Siccar Point that day. 'Whilst we listened with earnestness and admiration to the philosopher who was now unfolding to us the order and series of these wonderful events, we became sensible how much further reason may sometimes go than imagination may venture to follow.'[27]

It would be one of geology's most transformational contributions to human thought, allowing us to 'burst the limits of time', as one eminent scientist later put it.[28] Time, according to Hutton, had 'no vestige of a beginning, no prospect of an end'.[29]

The discovery of the future

It was around this period that Western societies went through a profound change, which some historians have called the 'discovery of the future'.[30] In roughly the same decades that Hutton was deepening Earth's history, various European intellectuals and writers were also looking forward, embracing the idea that time extended far ahead of them.

In 1755, the philosopher Immanuel Kant wrote that 'millions and whole mountains of millions of centuries' lay ahead for humanity and nature, 'during which new worlds and new world systems will constantly develop and reach completion, one after the other'.[31] In this open, endless future, Kant saw new heights of enlightened civilisation, which he called 'den Ausgang des Menschen aus seiner selbst verschuldeten Unmündigkeit' – 'man's exit from his self-incurred immaturity'.[32]

Meanwhile, a longer view was emerging within the fiction of the period. In 1733, the Irish Anglican clergyman Samuel Madden published one of the first English works of futuristic fiction, *Memoirs of the Twentieth Century*. It was an epistolary novel featuring letters written by diplomats in the late 1990s, such as notes from the English ambassador to France reporting on quarrels between Louis XIX and the Vatican. The narrator, living in Madden's time, explains he has received the letters from a guardian angel (time-travelling machines were not yet imagined).

Before this period, writing about the future had been reserved for

astrologers or prophets. To be the author of 'chronica defuturo scribet' –
a chronicle of things before they are done – was to be associated with
folly and perversity.[33] Perhaps that taboo was partly why Madden had a
change of heart, publishing anonymously and then seeking to suppress
distribution of copies.

While *Memoirs of the Twentieth Century* was a work of satire, rather than
prediction, it was an innovative piece of writing. It has been compared to
Gulliver's Travels, published seven years earlier, which used distant lands
as a way to satirise English society of the period. Madden, by contrast,
used distant time. That said, Madden was not a great writer, described
by some as tiresome, incoherent and unsophisticated in his satirical jabs.[34]

A few decades later, in 1770, Louis-Sébastien Mercier published *L'An
2440*, a utopian novel about a man who sleeps for hundreds of years before
waking up to navigate an idealised Paris of the twenty-fifth century. It
was a device Mercier used to highlight the perceived shortcomings of the
society he lived in: his protagonist discovers a secular, pacifist France of
the future, where there's no war, slavery or vice. Far more popular than
Madden's book, Mercier's would be published in at least twenty editions,
with more than 60,000 copies sold in various languages. Since he portrayed
a future France without religion, the Christian Church placed it on its
index of banned books. In Spain, where it was deemed heretical, the King
supposedly burned it himself.[35]

These eighteenth-century ideals about the deep future were not to
last. The French Revolution and other political changes in Europe would
nudge many utopian, future-facing intellectuals and writers into a more
pessimistic and circumspect mode.

Kant's views on the far future, for example, would evolve as he grew
older and the end of the century approached. He became increasingly
preoccupied by the possibility that humans could become extinct.[36] So,
while the future was 'discovered' during this period, it was accompanied
by the realisation that human beings might not always be in it.

Meanwhile, among ordinary people, living their daily lives, ideas about
long-term time were still unevenly distributed, to say the least.

The philosopher Ernest Gellner, who argued that each society has its own view of time, illustrated this contrast with an anecdote about two illiterate Swiss peasants from the Taugwalder family – a father and son, who were both called Peter.

In the mid-eighteenth century, the Taugwalders had survived the disastrous first attempt to climb the Matterhorn, which killed four English men. Afterward, they would tell the story of their roles as guides, and what happened. One of the climbers had slipped, pulling three more off the rock. Taugwalder-the-father had tried to hold on to them with the safety rope, but the hemp snapped, and they fell to their deaths.

When Taugwalder-the-son became old himself, he sometimes became confused about his identity, thinking that *he* had been his father in the story, clinging to the rope. Some put it down to senility, but Gellner didn't agree. 'After all, in the line of Taugwalders, there had always been fathers (with beards) and sons (without beards),' he suggested. 'At the time of the adventure, there was an old one, with beard, and a young one, without beard. It would have been absurd for him to identify with the young beardless one.'

In the society of the Taugwalders, children grew up to have exactly the same life as their parents. In such a community, there was still no sense of progress, and like the time of cathedral-building, the future was no different to the present or the past.

Evolution and industry

However, by the nineteenth century, the awareness of nature's deeper timescales was growing among the elite and educated. In 1832, the writer Thomas Carlyle made one of the first known references to 'deep time', albeit in a literary rather than geological context, speculating whether the work of the English writer Samuel Johnson would last the ages.[37]

The scientific and intellectual lengthening of past and future during this period was also paving the way for great strides forward in our understanding of ourselves and the planet. In particular, it allowed Darwin to propose his theory of evolution, bringing a profound reading of natural-world time that didn't place us at the centre. In 1859's *On the Origin of Species*, he wrote that 'the whole history of the world as at present known, although of a length quite incomprehensible by us, will hereafter be recognised as a mere fragment of time, compared with the ages which have elapsed since the first creature, the progenitor of innumerable extinct and living descendants, was created. In the distant future I see open fields for far more important researches.'

Meanwhile, the astronomer William Herschel had realised that deep time had been written into the night sky all along. In old age, he told an acquaintance: 'I have looked further into space than any human being did before me. I have observed stars of which the light, it can be proved, must take two million years to reach the Earth.'[38] His acquaintance, the poet Thomas Campbell, apparently had something of a temporal revelation during the conversation: 'I really and unfeignedly felt at this moment as if I had been conversing with a supernatural intelligence,' the poet wrote. 'After leaving Herschel, I felt elevated and overcome . . . [they were] some of the most interesting moments of my life.'

In the nineteenth century, clocks were also ever-more common. While in large parts of the world, including France, Asia and Islamic nations, people did not recognise that 1800 had begun because they followed different calendars, a sense of shared timekeeping day-to-day was emerging across Europe.

The days when churches were the main arbiter of hours and minutes were now long gone, and the West was entering the era of 'industrial time'. If there was anyone left who – like the Ancient Roman parasite (see page 15) – wished to live totally by their own internal tempo, they would have struggled. And amid this acceleration of progress, children grew up to have radically different lives to their parents – the days of Taugwalder-the-father and Taugwalder-the-son were beginning to fade away.

Historians argue that the machine that drove the Industrial Revolution was not the steam engine, but the clock, because clocks could synchronise people.[39] The advent of industry brought an intensification and commoditisation with far-reaching effects on Western attitudes towards the future.

The sociologist Barbara Adam once argued that industrial time is one of several different 'timescapes' that have shaped modernity.[40] Like cityscapes or landscapes that describe geographic space, timescapes describe temporal realms: natural time, psychological time, religious time and so on. According to Adam, what made the industrial timescape so powerful is that it introduced its own demands on the future. Industry commodified time, making it into a 'quantifiable resource that is open to manipulation, management and control', she writes. And as well as synchronising people, aiming to boost their efficiency and economic value, industrialisation opened up the possibility that the future itself could be exploited.

To take an example, consider that agriculture before the Industrial Revolution had been predominantly shaped by the natural timescapes of climate and seasons. In the mid-1800s, however, British scientist-entrepreneurs discovered that crushed bones treated with sulphuric acid vastly improved crop yields.[41] Coupled with the other impacts of industrialisation, these 'superphosphate' fertilisers would impose a new timescape on farming, meaning farmers no longer had to rely on the capricious cycles of nature. Crucially, it was a phase-change that came at the expense of the future, via phosphate pollution and soil depletion.

In sum, at this point in history, the future became a space that could be exploited on industrial scales to serve the economic needs of the present. And in many ways, this has been the story ever since.

Utopia returns

By the late 1800s and early 1900s, the impact of industrialisation on Western culture and its sense of time had fully taken hold. But while accelerated commerce was nudging societies towards a focus on present-day profits, the discoveries of science and technology had also helped to reignite a sense of the long view. It was an age of flight, radioactivity, electrification and motorcars. The changes brought by these innovations helped to open up an optimistic version of the long view, particularly oriented towards the future.

Artists and architects began reflecting those extending horizons through movements such as futurism and modernism. 'To be "new" was a value in itself,' writes the historian Hölscher.[42] 'It was not associated with dangerous alteration, but with progress, advantage in the competition of daily survival. It became common wisdom that natural order dictated that man had to change in order to be fit for the future. By the turn of the century, the "new man", the "new woman", and the "new society" were synonyms for their ideal forms.'

Echoing the utopian ideals of Mercier's L'An 2440, in 1888 Edward Bellamy published Looking Backward, 2000–1887, in which a man awakes to find himself in Boston in the year 2000, in which inequality in education, healthcare, wealth and social status no longer exist.

A few years later in 1895, H. G. Wells placed readers in a deep future rarely dreamed of before. In his novel The Time Machine, his protagonist travels hundreds of thousands of years into the future to 802701, where he meets the elegant Eloi race and the primitive Morlocks. While Wells's vision was hardly optimistic, it deployed staggeringly long timescales. Towards the end of the story, the Traveller jumps ahead again by thirty million years to observe the last living creatures on Earth – menacing giant crabs – and then again until he observes the death of the Sun and a freezing, lifeless planet.

In 1902, Wells gave a talk at the Royal Institution in London offering his thoughts about the long view, which he described as a minority – but burgeoning – perspective.[43] He argued there are two types of mind. One he disparagingly framed as past-oriented, and held by the majority of people of the age. 'The predominant type, the type of the majority of living people, is that which seems scarcely to think of the future at all, which regards it as a sort of blank non-existence upon which the advancing present will presently write events,' he said.

'The second type, which I think is a more modern and much less abundant type of mind, thinks constantly and by preference of things to come, and of present things mainly in relation to the results that must arise from them.' The first is 'retrospective in habit', associated with the 'passive', and 'legal and submissive' type, who is defined by precedent; the second is 'constructive in habit', and is 'legislative, creative, organising, or masterful, because it is perpetually attacking and altering the established order of things'.

It's not too hard to infer which type Wells thought described himself.

Around this time, science would reveal that the magnitude of past and future was far greater than they had realised. In the late 1800s, geologists had believed our planet was somewhere between 20–100 million years old. They based these figures on calculations such as the crust's cooling rate and, like Hutton did with his unconformity, making stratigraphic comparisons between rock formations. While there was disagreement over the exact number, it seemed that the next step was simply to refine the figure. However, the arrival of radiometric dating in the early 1900s revised that age upward by an overwhelming margin – geologists had been wrong by *billions* of years.

The discovery of radioactivity, meanwhile, would massively extend people's horizons into the deep future. Whereas Victorians like William Thomson (later Lord Kelvin) had plotted a lifespan for humanity on the order of 300,000 years – based on his projections for the Sun's death – the realisation that the atom holds vast stores of energy also opened up the potential for a vast tomorrow, according to historian Thomas Moynihan.[44]

When scientists realised that an atomic-powered Sun had billions more years left to burn, it 'completely inverted the ratio of expected future to established past. From thinking they lived near history's end, people now recognised they could be living during its very beginning. Humanity's universe, no longer decrepit, now seemed positively youthful,' writes Moynihan. As the physicist James Jeans observed in the 1920s, human beings may, in fact, be 'creatures of the dawn', with 'unimaginable opportunities for accomplishment' ahead.

Wells was similarly awestruck by the scientific developments of the day, writing in 1922: 'What man has done, the little triumphs of his present state, and all the history we have told, form but the prelude to the things that man has yet to do.'[45]

So, when visitors to Hutton's Unconformity and Enlightenment thinkers had imagined the 'abyss of time' in the 1700s, they really had no idea how deep it would turn out to be. For the first time, humanity was experiencing a taste of its true insignificance within the grand story of the planet and Universe – and it was dizzying. Outside science, however, this long-term view would not be universally embraced. It would also take a darker turn.

A darker future

In the subsequent decades of the twentieth century, many of the cultural norms and habits that shape the West's failure to take the long view today were beginning to form – particularly within the worlds of capitalism, politics, media and technology. We'll explore those specific changes in more detail in later chapters, but on the bird's-eye scale we are taking now, the twentieth century for the West can be framed as a dizzying series of contractions and reboundings. It was a time of acceleration, ideological upheaval and rapid cycles of pessimism and optimism, much of which laid the foundations for what is happening in the world today.

This was, after all, a century of global warfare during which cultural perceptions of time would be shortened by the salience of conflict. In the same way that the French Revolution dampened the future-facing optimism of the 1700s, the human tolls of the world wars would inevitably influence the emerging long-view perspectives of the time. During times of crisis, the present almost always takes priority.

And while the 1920s brought an economic boom and sense of possibility that opened up a long view in some ways, it closed it in others, by establishing capitalist norms that would help to shorten horizons later (we'll look specifically at how in Chapter 2). As the 1930s approached, the Great Depression had plunged the world into pessimism again, and the Second World War followed.

By the 1940s, if there remained a sense of a longer future, it took a darker tone – at its most extreme in the totalitarianism of the Nazis. This was a long view defined by destruction and horror, with a willingness to throw out the past (and much more) in order to make way for a 1,000-Year Reich.

In 1941, George Orwell lamented how the technology-centred utopianism evoked by H. G. Wells had been fully appropriated by fascism. 'Much of what Wells has imagined and worked for is physically there in Nazi Germany. The order, the planning, the State encouragement of science, the steel, the concrete, the aeroplanes, are all there,' he wrote. 'Creatures out of the Dark Ages have come marching into the present, and if they are ghosts they are at any rate ghosts which need a strong magic to lay them.'[46]

The Nazis had a predilection for technology, speed and modernism, and political propaganda became entwined with engineering. In the late 1930s, the party launched 'Voyages of Technology', which were touring train-based exhibitions to showcase German technical achievements. Led by Fritz Todt, a construction engineer and senior party official, the Voyages showed off futuristic materials such as fibreglass and synthetic rubber, with exhibits about the Autobahn and other aspects of cutting-edge Deutsche technik.[47] Prizes of washing machines, radios and refrigerators were handed out, along with pamphlets promising easier work, better efficiency and the potential to bring 'the fulfilment of wishes and dreams closer'.

The goal was to persuade citizens that Nazi engineering held the key to reducing unemployment and boosting industry – but at the same time, the propaganda blamed Jewish factory-owners for economic hardship. In one town, the engineers on the train-tour came across a 300-year-old Jewish monument – they got drunk and painted over the Hebrew inscription with black, weather-resistant varnish, before beating up locals.[48]

As the Reich grew, dreams of an 'ideal man' in the future begat cruelty in the present, as scientists experimented on those judged to be inferior, such as the Jews, Roma or mentally ill. The ideas of eugenics came to the fore, with plans to re-engineer the genetic make-up of humanity over the long term.

As the historian Charles S. Maier once wrote, 'whereas the nineteenth century served as the tablet on which historians inscribed stories of progress, the twentieth century story emphasizes narratives of moral atrocity or moral struggle'.[49]

Post-Second World War, technology's more utopian promises briefly returned in the West. The enthusiastic embrace of nuclear power, chemistry and automation brought a return to a lengthier, optimistic sense of tomorrow, particularly in the US. The late 1950s and 60s have been described as a Golden Age of American Futurism.[50] The Jetsons cartoon aired, NASA was founded and magazines published articles about the life of leisure and freedom Americans would soon all experience, powered by robots, flying cars and jetpacks.

But that attitude soured and faded in subsequent decades as the promised utopian future failed to arrive. At the time, Graham Swift lamented in his novel Waterland: 'Once upon a time, in the bright sixties, there was plenty of future on offer.'[51] Less so as Western economies stagnated, the conflict of the Cold War played out and awareness of environmental degradation grew.

So, if there was a long view in the West during the twentieth century, it was built on the scientific discoveries of the deep past but was particularly oriented to a future marked by technological progress and reinvention. Occasionally, this future even held utopian glamour. However, these optimistic hopes have dissipated over the past few decades.

As the German cultural historian Aleida Assmann writes: 'Experience teaches that not only specific visions of the future have crumbled, but that even the concept of the future as such has changed beyond recognition. In many areas such as politics, society and environment, the future has lost its lure. It can no longer be used indiscriminately as the vanishing point of wishes, goals and projections.'[52] Assmann makes the case that the West can no longer look to the future while also reckoning with the memories of traumas and consequences of its recent acts of the past, from conflict to colonisation. Drawing on Hamlet, she asks: 'is time out of joint?'[53]

The time-blinkered age

As the final few decades of the twentieth century arrived, the seeds had been sown for a widespread transformation that persists today. The West was entering an age when short-termism began to seep invisibly into many realms of life. The future didn't disappear exactly, but the needs and demands of the present became louder and more dominant, holding sway over politics, business, media, popular culture and much more.

These changes were gradual, fragmented, and difficult to blame on any single factor. But if you had to choose one example that captures how this attitude had taken root, you might well land on the case of the I-35W bridge in Minneapolis, and the disaster that happened there one summer evening during rush hour. It happened in seconds, but was decades in the making.

It was just after 6 p.m. on 1 August, and commuters were travelling home on an uncomfortably humid day. The temperature had topped 30°C (86°F), and the interstate highway I-35W was filled with cars, buses and trucks.

Among them was Kimberly Brown, travelling in a silver Saturn car with her friend on the way to a soccer game. Heading north, they moved through traffic as they crossed the bridge over the Mississippi, which bisects Minneapolis early on in its 3,730km (2,318 miles) route down to the Gulf of Mexico.

Suddenly, there was a loud sound of cracking metal. 'Then I feel the road shake,' Brown writes in her 2018 book about what happened that day.[54] The highway surface rippled and bounced. 'Again, stronger . . . Rumbles turn to waves, and the flat bridge deck breaks.' Everything dropped: the road, steel girders, and all the vehicles plummeted 30 metres (98 feet) down towards the river. 'The falling world doesn't just clatter – it booms,' she recalls. 'The concrete fractures. Steel scrapes and roars.'

The whole eight-lane, 580-metre (1,903-feet) I-35W bridge had collapsed, filling the air above Minneapolis with a plume of smoke and dust. Afterward, some spoke of moments of complete silence – and then a different kind of cacophony as the sirens of emergency services broke out, and media helicopters hovered above.

Unlike those driving other vehicles, Brown and her friend did not end up in the water. After it landed, the car ended up awkwardly suspended at a 45-degree angle across two broken slabs, with steel reinforcements poking out, only a few metres from the river flowing around them. Panicked, they pulled themselves out through the driver's side window. Around them, others struggled to escape from flooded or crushed cars, and a truck started to set on fire. Eventually, they were picked up by a boat and taken to safety.

That day in 2007, thirteen people died and 145 were injured, in what was one of the worst bridge disasters in US history. The senator Amy Klobuchar – a future Democratic presidential candidate – commented at the time that 'in America, a bridge should not simply fall down'. If anything, it was a statement of bewilderment that such a disaster could have happened.

Yet when engineers and researchers later investigated, they uncovered that it was entirely possible to have foreseen the bridge disaster. Built in the 1960s, the bridge's designers had used thin 'gusset plates' to connect the bridge's steel girders. These flat plates are bolted over joints where two girders meet, and look a bit like the web between a duck's toes. It was a flawed design, but it kept costs down, and public officials were not incentivised to go into anything too deeply. What was particularly extraordinary was that inspectors had been labelling the bridge 'structurally deficient' since 1991. If they had attached strain gauges, they would have seen the stress on the plates increasing over time. Especially when new road lanes on the bridge added extra weight. On the day of collapse, heavy construction vehicles and materials tipped the design over the edge. The gusset plates snapped, and there was no redundancy. When they failed, the entire bridge fell.

The Tenth Avenue bridge, adjacent to the I-35W, shows what a less efficient, but more resilient, design could have looked like. Built in 1929, it has independent arches separated by concrete pylons that divide the structure into discrete sections. The columns might seem oversized for the load they carry, but even if one of the columns or arches failed, the damage would be isolated. In the 1960s, such redundancy might have seemed unnecessarily over-engineered, expensive and wasteful.

The design of I-35W was far from isolated – around 465 bridges in the US had similar designs when it fell, and up to 60,000 bridges in the US are still considered structurally deficient.

So the culprit, in engineering terms, was a piece of metal that failed. But the real source of blame was more diffuse than that, and spoke to a time-blinkered culture that ignored how decisions in the present could affect people in the future.

Thomas Fisher, a professor of architecture at the University of Minnesota (which has a campus close to the site of the disaster), sees the downed I-35W bridge as a metaphor for what he calls a 'fracture-critical' society.[55] In such a society, companies, regulators and politicians all make short-sighted decisions that suit their present-day needs, which then combine to cause catastrophes down the line.

And faulty bridges are just the tip of the iceberg. All across the US, people are now facing the consequences of ageing electrical grids, sewer pipes, water supply, fibre-optic networks, transport infrastructure and flooding prevention. Every so often an event happens that exposes lack of investment: a state-wide power outage that leaves millions without heat, a hurricane that floods river levees, or a train that comes off the rails. The media and politicians pile in, and promises are made to fix the damage and find justice for the victims – but then attention moves on, and the underlying long-term causes are ignored.

According to the economist John Kenneth Galbraith, after the Second World War the US became a nation that accepted 'private affluence' and 'public squalor' as the norm. 'That put designers and their governmental clients under increasing pressure to do as much as they could as efficiently as possible, given the relative lack of money,' writes Fisher. 'The paradox is that, at the very moment that America could have afforded the very best infrastructure in the world, we decided instead to direct far more wealth into private hands and to begin to squeeze the public realm of funds.' What didn't help was a technological hubris boosted by winning the Second World War, he continues.

'The I-35W bridge stands not just as a physical structure over the Mississippi River but also as a political, economic, and social symbol of postwar America,' Fisher concludes. 'Shortsighted efficiency at the start can lead to extraordinary costs in the end.'

If it was just a question of problematic infrastructure alone, it might be possible to tackle with enough money. But this is a problem that underlies so much more: it spans all realms of modern life, and not just in the US. It is a short-termist outlook that has spread all around the world.

You can see it in business, where quarterly reporting encourages CEOs to prioritise short-term investor satisfaction over long-term prosperity. You can see it in populist politics, where leaders are more focused on the next election and appeasing their base than the long term. And you can see it in our collective failure to tackle slow, creeping societal problems that play out over decades or more, from climate change to inequality. None of these trends is brand new, but their influence has grown to reach further across time than ever before.

Every so often, like the I-35W bridge disaster, the fallout happens and becomes impossible to ignore. It might be a burning oil rig pouring millions of gallons of oil into the ocean, a profit-driven housing bubble triggering a global financial crash, a record-breaking fire or flood destroying a neigh-bourhood, or a pandemic we knew was coming killing millions around the world. But then another crisis comes along, and attention moves on.[56]

Zooming out, the historian François Hartog has highlighted three periods of time across Western civilisation to date, which he calls 'regimes of historicity'.[57] Up until the 1700s, he says, only the past informed the present. In the second regime, over the ensuing 200 years, the present was conceived of in terms of the future. This all roughly lines up with our journey through the ages so far.

But sometime around the late 1980s, Hartog argues Western civilisation fully entered the third regime, when a series of societal trends converged to begin a temporal state that he calls 'presentism'.[58] He defines this period as 'the sense that only the present exists, a present characterised at once by the tyranny of the instant and by the treadmill of an unending now'.

Towards the end of the twentieth century, as the Berlin Wall came down – around the same time that the historian Francis Fukuyama was predicting the 'end of history' and the enduring dominance of liberal democracy – Hartog says the West fully abandoned the idea of future-oriented modernity. Around this period, he says, Western nations embraced 'un présent monstre': a 'monstrous present'. Today, 'the future is not a radiant horizon guiding our advancing steps, but rather a line of shadow drawing closer', he writes.

Other scholars have converged on similar conclusions,[59] describing the present as 'tyrannous', 'ever-broadening', 'omnipresent', 'extended', and 'all-pervasive'.[60] We are now living in a time, they say, when the past is no longer a source of wisdom or succour, the present has become all-present and the future is not what it used to be.

With an eye predominantly on the capitalist culture of the West, the historian-anthropologist Jérôme Baschet writes of the tyranny of a perpetual present, colouring all views of time with its priorities: 'The deified today, the triumphant oblivion, and the present drawn out into eternity are but three ways to express the same reality: in the dominant time of the globalized market, there is no longer any past to learn to know, any future to look forward to.'[61]

Unlike our ancestors, we are fully aware that tomorrow will be different to today, and that our societies exist within deep time. Unfortunately, that hard-won knowledge is ignored in practice. In a time-blinkered age, all modes of thinking are shaped primarily by present-day concerns, meaning the long view is often seen only through the lens of satisfying current needs, increasing profit or winning political battles. Or as the psychologist Daniel Gilbert once put it: 'If the present lightly colours our remembered pasts, it thoroughly infuses our imagined futures.'[62]

There are multiple forces fostering this time-blinkered mindset in Western culture. Some point to that often-blamed scourge, the internet. Others lament the intersection of twenty-four-hour news media and politics, which encourages decision-makers to focus more on headlines or polling than future generations. Hartog blames the capitalist, consumerist norms that came to dominate Western culture by the late twentieth century. During this period, 'technological progress kept forging ahead, and the consumer society grew and grew, and with it the category of the present, which this society targeted and, to an extent, appropriated as its particular trademark', he writes. 'Soon came the supremely imperious time of globalisation in the form of a world economy, which pushed for ever greater mobility and referred increasingly to "real time".'[63]

However, as with many ailments, there is no single cause of the time-blinkered age, rather the convergence of many – both cultural and psychological. I call these *temporal stresses* and *temporal habits*, and we'll explore them in the chapters to come.

Crucially, there is no need to despair about this predicament, nor mourn the loss of the future entirely. If our account of temporal history and diagnosis of Western society is correct, then short-term thinking is an emergent property of the cultural, economic and technological moment, which is not necessarily locked in forever, nor totally out of our control.

Given that our ancestors have made great leaps forward in their perceptions and awareness of their place in time, there is no reason to assume that it cannot happen again. If the West can learn to identify the sources of its predicament, then achieving a more sophisticated long view of myriad possible futures is not impossible.

After all, a deeper view of time has only just been discovered. The ability of *Homo sapiens* to look far beyond our present is a relatively new evolutionary invention on this planet, so we may still only be early in our development towards a richer understanding of our role in the long term. Our descendants may one day look back on the lack of awareness and foresight today as we look back at the flatter perspectives of our ancient ancestors.

If we are to achieve the long view, however, first we must build a more in-depth picture of the reasons why the time-blinkered age began, and the hidden temporal stresses at its core. Until we fully understand the problem facing contemporary societies, we cannot lengthen our perspective.

So, over the next two chapters, let's dive a little deeper into two major sources of time-blinkered thinking in modern life – Western-style capitalism and politics – motivated by the logic that understanding the incentives and deterrents of these realms can reveal the hidden short-termism within them, and elsewhere. The good news is that there are a few rare individuals and organisations in business and politics who have managed to transcend the pressures they face to adopt the long view – and there are lessons that we all can learn from their experiences and wisdom.

2

Selling Short:
The Unforgiving Immediacy of Capitalism

'There is no man perhaps, to whom a good to be enjoyed today, would not seem of very different importance, from one exactly similar to be enjoyed twelve years hence, even though the arrival of both were equally certain.'

JOHN RAE[1]

'The social object of skilled investment should be to defeat the dark forces of time and ignorance which envelop our future.'

JOHN MAYNARD KEYNES[2]

One day in the late nineteenth century, a German immigrant and entrepreneur called Henry Timken decided to test one of his inventions out on the streets of St Louis, Missouri. It was so impressive, it almost got a man arrested.

In American cities at the time, people relied on horse-drawn carriages to transport people and goods, but there was a limit to how much weight the animals could pull. Overloading carriages was deemed an offence in some states, after lawmakers began to prosecute animal cruelty in New York in the 1860s. One of the first cases, *People vs Tinsdale*, involved a driver and conductor guilty of tormenting their horses because the animals had been unable to pull a carriage full of passengers. (In Pennsylvania, the maximum fine for cruelly beating a horse was $200, but only $100 for abandoning your child under seven.[3])

The problem was not just weight, but friction. When bearing a load, the wheels of carriages become harder to turn. Timken realised it could be possible to make heavier carriages travel faster and further – but without harming the horses. 'The man who could devise something which would reduce friction fundamentally would achieve something of real value to the world,' he would say later.[4]

With a colleague, Timken started experimenting with a new design for roller bearings, small rotating cylinders that could be placed inside the wheels of carriages. Satisfied with an experimental design, the pair sent a wagon and horse equipped with their handmade bearings on to the streets of St Louis, the fourth largest US city at the time. The large spoked wheels bounced over cobbles and dirt roads, as the wagon zipped passed drugstores and food markets.

Timken's experiment came to an abrupt end, however, when the driver was stopped by police. The load was too large, they explained, and cruel to his mule. It was only smoothed out when his son, who was the owner of the wagon, appeared in court to explain how the roller design worked – and successfully avoided a fine.[5]

In 1899, Timken founded the Timken Roller Bearing Axle Company to exploit his invention – and it would go on to become one of the US's most successful international family businesses for decades to follow.

His smart design for reducing friction would also find uses far beyond horse and carriages. A few years later, he moved the business to the small town of Canton, Ohio, to be closer to the promising automotive industry in Detroit. When the company started to manufacture its own high-quality steel too, it was perfectly positioned to take advantage of the growing demand for motorcars, and later vehicles and weaponry for the world wars. The Timken Company grew into a thriving cor-poration in Canton, keeping thousands of workers employed as other manufacturing jobs disappeared, and passing between five generations of Timkens.

Fast-forward to the twenty-first century, and the company had passed into the hands Tim Timken, the great-great-grandson of Henry. The company had been loyal to the people of Canton: base salaries were an average $23/hour, higher than rivals', and the firm was widely admired for donating to schools and museums in the area. It had also invested in giant one-of-a-kind mills that cost hundreds of millions of dollars, forsaking its short-term profits and shareholder returns for these capital investments. Despite recessions, the strategy had paid off in the long run, allowing Timken to keep pace with technologically advanced competitors overseas, while other manufacturing industries in rural America had withered.

What happened next, however, serves as an illustration of how the temporal stresses embedded within modern capitalism encourage a shorter-term view. Tim Timken and his colleagues would face a fight over the fate of the company unlike any they had encountered in its long history. But what made the episode striking was that there were no bad guys – at least, not in the traditional movie sense. As we'll discover, what made it happen was systemic.

One day, the analysts at a California firm called Relational Investors had noticed something about Timken that got their attention. They had been scouring the market for undervalued share prices, and concluded that they could boost prices if the century-old company was split into two. Henry Timken's company may have reduced friction nationwide 100 years ago, but now the firm itself was deemed to be less efficient than it could be.

Relational's analysis suggested that two Timken companies would become more appealing to buyouts by a bigger company. It also suggested that the two parts could take on more than $1 billion in new debt.

Relational worked up a plan. A minority investor can't just waltz in and demand that a century-old company break up. So, over a number of months, their traders began to buy shares, acquiring a bigger and bigger stake in Timken. Eventually, this gave them clout.

What followed was a bruising public fight over the fate of Timken.[6] To break up the company, Relational needed to get other shareholders on board, so they launched a sophisticated campaign, with compelling presentations and websites like unlocktimken.com, arguing that the reasoning for maintaining the status quo was based on 'amorphous arguments and faulty math'.[7] The Timken family didn't support Relational's plan, so played defence by launching their own site timkendrivesvalue.com, pleading with shareholders to keep the company together (at the time of writing, this URL was dead and, aptly, redirected to a eulogy writing service). Meanwhile, workers and their families in the community grew worried that new owners could cut jobs and move production elsewhere.

If it was a bad movie, the director might frame it as 'greedy investor' versus 'valiant small-town manufacturer'. But in reality, it wasn't that simple. By then, Timken was a major corporation whose executives were paid multimillion-dollar salaries with powerful connections in the Republican Party. And a charitable view of Relational was that it was just trying to maximise its own bottom line, while facing its own pressures. According to the *Wall Street Journal*, Relational's founders – one of whom was fighting throat cancer at the time – preferred to cooperate with CEOs rather than fight them.[8]

Crucially, there was also a third actor. Early on, Relational had savvily brought another major shareholder on board to support its plan: the California State Teachers' Retirement System, one of the world's largest educator-only pension funds. This made it a more nuanced argument. It became a fight between the needs of an Ohio company and its community, and the needs of almost a million teachers and their families.

It was all too much for Timken to withstand. Relational won, and the Timken company was split up. The bearings part of the company, which Henry had started in 1899, was taken over by new management from outside the family. The board swiftly cut pension fund contributions and halved capital spending.

The other new company, Timken Steel, carried on with Tim at the helm, attempting to continue the family legacy through steel-making alone. It got off to a strong start – but a few years later, the value of the company had plummeted as steel demand tumbled. In 2019, the workforce was cut by 14 per cent, college scholarships for employee children were stopped, and Tim was forced out as CEO.[9] He went on to run a political lobbying firm.

Relational never got involved in all that though. A few months after the split, they had unloaded shares and moved on. Later, the fund dissolved.

The Timken businesses continued to exist, split between the two companies – but it's more of a brand than a family business. The legacy that Henry Timken built at the turn of the twentieth century is effectively over, the five generations of custodianship ended by a relatively fleeting fight over share prices. Of course, it's possible that a family-run Timken company would have gone on to face financial difficulties even if Relational had never targeted the firm, but it's an alternative history we'll never know.

When Henry Timken founded his company more than 120 years ago, he probably could never have imagined that his family's legacy would play out in this way. A piece of advice that he once gave to his sons takes on a new light after all that happened.[10] 'To be successful, you must be independent,' he said. 'If you want to lead in any line, you must bring to it independence of thought, unfailing industry, aggression, and indomitable purpose. If you have an idea that you think is right, push it to a finish. Don't let anyone else influence you against it. If we all thought the same way, there would be no progress.'

That is all very well, but the world that Henry's descendants live in has changed. It is one where a responsibility passed between generations can be lost because an analyst in California needs to boost the numbers on their screen, and where the needs of a family and community can be pitted against teachers and their pensions. There are no malign actors here: instead, everyone is operating within a system shaped by time-blinkered norms, incentivised to prioritise short-term gains over long-term stewardship.

Stories like Timken's are far from isolated. In 2019, there were more campaigns by new activist investors than ever before, and such behaviour is now 'totally destigmatised', according to the financial advisory firm

Lazard.[11] Meanwhile, professional investors are buying and selling their stakes in companies at a faster rate than at any time in the past century. On the New York Stock Exchange, the average holding time for shares in the 1960s was around eight years. These days, it is only a matter of months.[12]

It's therefore perhaps unsurprising that the average age of a company listed on the S&P 500 – the 500 largest US publicly traded companies – fell from almost sixty years in the 1950s to around twenty years today.[13] It's projected to be around fifteen years by 2027.[14] Some of this churn may have happened due to the dizzying valuations of twenty-first-century technology companies, which has knocked slower-growing companies off the list, and some of it can be explained by mergers. But it remains true that the Western corporation tends to have a short life compared with organisations in other nations and sectors. They could in principle last for centuries, but on average barely last a few decades.

'The high corporate mortality rate seems unnatural,' the former Shell executive Arie de Geus once wrote.[15] 'No living species suffers from such a discrepancy between its maximum life expectancy and the average span it realises. And few other types of institution – churches, armies, or universities – have the abysmal record of the corporation.

'If you look at them in light of what they could be, most commercial corporations are underachievers. They exist at an early stage of evolution; they develop and exploit only a small fraction of their potential.'

Inside corporations, short-termism is becoming ever-more common, according to research organisations who track short- and long-term behaviours in business.[16] Over the past two decades, the Boston-based organisation FCLT Global (Focusing Capital on the Long Term), has observed a steady decrease in long-term habits within global companies, such as fixed investments, and an increase in short-term habits such as share buybacks and dividends.[17]

A litany of invisible cultural norms and practices embedded in twenty-first-century capitalism are pressuring companies, investors and individuals into blinkered decisions, but crucially no single actor can be blamed. To understand how these temporal stresses came to dominate modern capitalism – particularly in the West – you have to zoom out and consider how all the parts interact: shareholders, companies, regulators, fund managers, legislators and more. Capitalism is not necessarily inherently time-blinkered, but over the past century various actors have invented and introduced practices, incentives and deterrents that, collectively, have discouraged a long-minded view.

This matters not only for business. Capitalism's unforgiving immediacy influences the world far beyond the worlds of shares, boardrooms and trading floors. Its habits are entwined with politics and broader society, and steer the world's response to the grand challenges of the century, from climate change to public health crises.

The good news is that there are various companies and organisations around the world who, despite facing shortening temporal stresses, have successfully managed to look longer term. Capitalism in the twenty-first century may force a focus on the present – but as we'll discover, it can be changed. While its habits may be entrenched and far-reaching, they are not necessarily irreversible.

But first, we need to understand the origins of the systemic temporal stresses shaping the world of business. And we'll begin by examining some highly influential practices that took root in the early twentieth century.

Beating the gun

Around the same time that the Timken company was flourishing in the US, the British economist John Maynard Keynes was writing about where he believed capitalism was headed in the long run, as well as the behaviours he observed on the financial markets back then.

One of the most important economists of the century, Keynes is perhaps best known for his influence on how governments should respond to economic crises, but his other observations give a sense of how the future looked in the 1920s and 30s. Some things he would definitely get wrong; others were remarkably prescient.

Keynes was a member of the Bloomsbury Group, a collection of intellectuals, writers and artists including Virginia Woolf, E. M. Forster and
his early lover, the painter Duncan Grant. At the time, Woolf described
Keynes as a 'truculent' and 'formidable' man, 'able to render any argument
that came his way with a blow of his paw, yet concealing, as the novel-
ists say, a kind and even simple heart under that immensely impressive
intellect'.[18]

It's possible the privileged bohemian lifestyle of Keynes's Bloomsbury
friends influenced how he imagined what business – and life more generally – would look like in the twenty-first century. In 1928, writing the first
drafts of a talk called 'Economic Possibilities for Our Grandchildren', he
predicted that productivity and growth in the West would lead to a new
era of leisure, well-being and moral enlightenment, and that, by the 2020s,
his grandchildren would be working for only three hours a day.

The love of money for money's sake – a 'somewhat disgusting mor-
bidity' – would be long forgotten, he wrote, and be seen as closer to a
mental disease. Sadly, external events would soon challenge that prediction. Shortly after Keynes began giving his long-view talk to small societies
in the UK, the global economy was in deep trouble.

The Roaring Twenties had been a tumultuous period, particularly in
the US. After the First World War, American wealth almost doubled, and
the New York Stock Exchange expanded significantly, fuelled by speculation from both professional traders and blue-collar workers putting their
savings into stocks. New financial inventions had allowed the average Joe
to invest in shares using borrowed money, which worked fine in the good
times, but would prove disastrously short-sighted when confidence waned.

On Black Tuesday, 29 October 1929, prices plummeted in the US
amid market panic, wiping out billions of dollars of investments. The
Great Depression had arrived. Banks collapsed, unemployment spread,
farmers' crops failed and homelessness grew. The rot spread around the
world, demonstrating the downside of a newly interconnected global
economy. As one historian has pointed out: 'The Wall Street crash of
1929 was the first economic event of truly global weight; producers and

consumers on every continent were reeling from its consequences within a few months.'[19]

As a deep recession endured, Keynes continued to cling to his optimistic view of the long run, predicting greater prosperity and leisure eventually. He republished updated versions of 'Economic Possibilities' as a more formal essay. But by the mid-1930s, he was also raising some of the first concerns that short-termism was taking root in financial markets. Unfortunately, it is this diagnosis that has proved more accurate with time.

In 1936, he observed a new trend in the behaviour of investors who he saw as taking a short view to make a quick profit. 'They are concerned, not with what an investment is really worth to a man who buys it "for keeps", but with what the market will value it at, under the influence of mass psychology, three months or a year hence,' he wrote.[20]

'The social object of skilled investment should be to defeat the dark forces of time and ignorance which envelop our future,' he argued, but what he was beginning to observe instead looked more like a game. 'The most skilled investment to-day is "to beat the gun", as the Americans so well express it, to outwit the crowd, and to pass the bad, or depreciating, half-crown to the other fellow.'

This wasn't necessarily 'wrong-headed', he wrote, rather the inevitable result of the way the market had been built to incentivise short-term speculation. 'It needs more intelligence to defeat the forces of time and our ignorance of the future than to beat the gun,' he concluded, but sadly the structure of the market was incentivising the latter.

As we know now, this behaviour would only become more common. Indeed, if Keynes had been alive in the twenty-first century, it would be intriguing to know what he would have thought about the Timken vs Relational fight and other cases where financial speculators saw a short-term opportunity to boost a share price at the expense of long-term stewardship. And what would he have made of algorithmic high-frequency trading, which allows traders to buy and offload shares in milliseconds?

We'll never know, but you might find clues in his own investment decisions. According to the journalist John F. Wasik, who wrote the

book *Keynes's Way to Wealth*, the economist managed money for King's
College, Cambridge, and two insurance companies, as well as friends and
family – making money in twelve out of eighteen years between 1928 and
1945. Notably, Keynes ignored the noise, seeing the daily ups and downs
of prices as an 'altogether excessive, and even absurd influence on the
market'. So Wasik speculates that if Keynes were around today, he would
probably ignore short-term price fluctuations and leave high-frequency
trading to the specialists. 'The moral of Keynesian investing is to play
the long game, stick to an investment plan and avoid being distracted,'
writes Wasik.[21]

Keynes would surely be disappointed, however, to learn that the life of
leisure he hoped for remains elusive. You and I are the 'grandchildren' that
he was imagining in the 1920s, but sadly that was a long view that never
happened. He might also ruefully observe the other temporal stresses
that nudged capitalist societies to become so time-blinkered, which he
didn't anticipate.

What were those stresses? One was a new 'invention' of financial
practice that emerged on the other side of the Atlantic around the same
time that Keynes was hanging out with the Bloomsbury Group. Over in
1920s New York, Western capitalism was slowly introducing a habit that
would curtail the long view of business in a profound way.

Quarterly thinking

In the early twentieth century, the New York Stock Exchange made a
seemingly reasonable request of its listed companies. The ask was simple:
every three months, companies would need to share financial details, projections and plans with the market, perhaps motivated by the idea that
transparency would discourage dishonesty. By 1931, 63 per cent of firms
had complied, and 17 per cent were reporting semi-annually.

Other US exchanges opposed it at first, and regulators were reluctant
to step in and make it mandatory, so this habit would take a few decades

to catch on more widely. But in the history of capitalist short-termism, the NYSE decision marked the genesis of a practice that would shape the world. This was 'quarterly reporting', a financial invention that would spread to other markets and have far-reaching consequences for business and societies a century later.

Slowly, quarterly reports began to be adopted as a norm among US firms, until a major turning point in the early 1970s when it really became embedded. After sporadic interventions in prior decades, the US Securities and Exchange Commission, which regulated markets, finally made the decision to make quarterly reporting, including earning projections, a requirement for all American firms. With this decision, the effects became supercharged.

Reporting regularly to investors might seem benign, but Arthur Kraft of City University in the UK, who has studied what happens when companies begin quarterly reporting, argues that it has had a profound effect on decision-making within companies. When senior leadership is forced to make promises to the market, so frequently, it's almost inevitable that this discourages the long view.

Studies by Kraft and other researchers suggest it has various tangible consequences.[22] 'I'm confident: when firms are forced to increase reporting frequency, they cut back on investments,' Kraft told me. Other studies suggest that quarterly reporting correlates with reductions in R&D spending, patents, advertising and hiring, as well cuts in discretionary spending and project delays. By one calculation made in 2020, such short-termism costs the 500 largest publicly traded companies almost $80 billion per year in foregone earnings.[23]

For these reasons, other countries have rejected quarterly reporting. After considering it, the European Union decided not to go ahead with introducing the practice.[24] And in the UK, regulators asked for quarterly 'Interim Management Statements' in 2007 but stopped it as a requirement seven years later.

Yet given the power of the US economy, and the interconnectedness of global business, there's arguably no escaping the influence of the quarter.

Though not mandatory, many large British companies release earning statements every three months anyway.

Zooming in to how it affects individual decision-makers and company leaders, Kraft blames two reasons for the blinkered behaviours encouraged by quarterly reports: '**discipline**' and '**myopia**'.

Discipline is where investors punish company leaders for longer-term management strategies. The Timken story would be one example, but you don't have to look far for more.

One striking case is that of Paul Polman, the former CEO of Unilever, who on the day he started the job announced his company would stop issuing quarterly reports. Instead, he said, Unilever would focus on long-term projects that might not pay off for years. He also demonstrated that commitment by pledging to decouple the company's growth from its environmental footprint,[25] and acquiring the sustainable cleaning brand Seventh Generation, inspired by the long-term principles of Native Americans (which we'll return to in Part III).

'Better decisions are being made,' he explained at the time. 'We don't have discussions about whether to postpone the launch of a brand by a month or two or not to invest capital, even if investing is the right thing to do, because of quarterly commitments.'

Polman's message to investors was clear. 'Unilever has been around for 100-plus years. We want to be around for several hundred more,' he said when he set out his vision.[26] 'So if you buy into this long-term value-creation model, which is equitable, which is shared, which is sustainable, then come and invest with us. If you don't buy into this, I respect you as a human being, but don't put your money in our company.'

It was a remarkable shot across the bows. 'I figured I couldn't be fired on my first day,' he later recalled.[27]

Unfortunately, the market was hardly supportive, punishing Polman through Unilever's share price, which plummeted after his announce-ment. It took years to change market attitudes, and the ride was bumpy. Reflecting on the episode years later, Polman said eventually Unilever did boost its long-term value, but only after the short-termist investors had sold and moved on, replaced by shareholders more willing to take longer bets.[28] He didn't change their minds, they simply moved on to companies that allowed them to beat the gun elsewhere.

Polman managed to weather the turbulence following his decision, but a lot of other company bosses would have been promptly replaced. Many leaders have lost their jobs when their share price falls precipitously. And many more may simply choose not to take a long-term decision that could lead to market disapproval and their firing – which leads on to the second cause of blinkered quarterly thinking identified by Kraft and colleagues.

If discipline is about external punishment, **myopia** is more self-inflicted. This is where company leaders pre-emptively take short-term decisions they believe will please the market, particularly to meet their earnings projections. It might involve cutting headcount, holding off on capital investment or rolling back training programmes, all to make the numbers on the balance sheet look better.

As one chief financial officer told researchers studying this behaviour, it's about avoiding a display of weakness and keeping up appearances. 'If you see one cockroach, you immediately assume that there are hundreds behind the walls, even though you may have no proof that this is the case,' they explained.

That CFO's comment was offered as part of a notorious survey of 400-plus senior US business leaders, which generated an astonishing statistic: nearly 80 per cent said that they were willing to take decisions that they *knew* would harm the company's value in order to deliver on their short-term earnings promises.[29]

'To me this was a stunner,' the corporate advisor Roger Martin wrote in a review of the findings.[30] 'I was not surprised that 80 per cent would do so – in fact, I suspect that the reality is closer to 100 per cent – but that 80 per cent would actually admit it.'

Worse, if senior executives have their pay tied to their company's quarterly or annual performance, then they may even personally benefit from making those short-term decisions. Chief executive bonuses tied to stock market performance were once uncommon, but by the turn of the twenty-first century almost half of CEOs were incentivised this way.[31]

Things have improved a little since, with greater longer-term incentives put in place for senior individuals, widespread acceptance that things need to change, and emerging efforts such as the launch of the Long-Term Stock Exchange in 2020. Still, during the Covid-19 pandemic, short-term pressures surged back for many business leaders, and the practices of past decades are deep-rooted and systemic.[32]

What's more, unhelpful targets are widespread at every level in present-day capitalist economies, and, as we'll discover next, these incentives can have profound effects on people's attitudes to the long view in business and beyond.

The tyranny of targets

If you're looking for moments where short-term business targets have proved disastrous, one particularly notorious case stands out.

In the 1960s, the car company Ford was facing challenges from overseas competitors producing cheaper, smaller vehicles, which were less costly to run. So Ford's chief Lee Iacocca announced a new goal for his company: to produce a car that would weigh less than a tonne and cost less than $2,000 to buy. His teams got to work. Iacocca had also set an ambition to get the car to market as quickly as possible.

The result was the Ford Pinto, launched in 1971, with a spacious coupé rear and elongated bonnet. It had been designed, manufactured and shipped in a little over two years, half the normal time. Within a few months, Ford sold more than 100,000 cars. Soon, however, a few unlucky drivers learned to their cost that the Pinto had serious problems.

One day, Lily Gray was travelling with her neighbour's thirteen-year-old son on the highway in California, when she stalled in the middle lane. The car was struck from behind, which triggered a fire. Gray was killed and the boy was seriously burned.

In the same year, three teenage girls from the Ulrich family stopped on a road in Indiana on their way to volleyball practice. Judy Ulrich, the driver, had noticed in her mirror that she had accidentally left the petrol cap on top of the car while filling up, so pulled up. Behind them, a van driver had dropped a cigarette on to the floor, so was reaching to grab it. By the time he looked up, it was too late, and he smashed into the Pinto. Again, a fire broke out. All three girls died.[33]

Later lawsuits would reveal that the Pinto had a deadly design flaw that had been missed in the rush to market. To meet Iacocca's targets, managers had signed off on safety checks that were never conducted – in particular on the fuel tank, which was squeezed in behind the rear axle. This positioning, exacerbated by the lack of crush space, made it particularly prone to setting on fire in a crash.[34]

As far as the engineers and managers were concerned, they had met the goals set by their CEO, but because speed and cost were prioritised over safety and reputation, the design flaws slipped through, and even when identified were not fixed. For the sake of meeting a target, people had died, and the fallout would damage Ford's image for years.

It's far from the only example of poorly framed targets incentivising people to take shortcuts and ignore the long view. In his influential book, *Moral Mazes* (1989), the sociologist Robert Jackall described a particularly problematic type of business person. This manager, he wrote, would have the uncanny ability to outrun his or her failures. They achieved this, Jackall argued, by 'milking the plant'.

It went like this. A manager would arrive at a plant or factory, and be handed a set of steep targets by the company board. They'd immediately crack the whip: asking more of their workers, and pushing the machinery harder than before. Productivity would rise. Months later, the targets would be hit and the board would be happy. Promotion or a new job soon followed, and the manager moved on. Left behind, however, was a mess. Workers were unhappy, the best talent had left for better conditions and the machinery has been run into the ground, needing expensive replacements. The next manager has to pick up the pieces.

If the person who 'milked the plant' sounds like someone you know, it's because the fast-turnover career opportunities of present-day business (and indeed many organisations) do little to root out such behaviour. People have often moved on to new roles before the consequences of their actions become accountable.

A different form of milking the plant can be observed in the world of finance. Here's one example that captured how bad targets underpinned the financial crisis of 2008, which was fuelled by Wall Street's short-sighted repackaging of unreliable sub-prime mortgages.

During a congressional investigation into the role of Wall Street in the crash, the US senator Carl Levin quoted a damning email conversation. As the timebomb of these bad loans brewed, a financial analyst had emailed a banker asking about what they were reading in a deal they were working on. Weren't the details of this problematic?

'IBG-YBG,' the banker replied.

When asked what that meant, the banker explained: 'I'll be gone, you'll be gone.'[35]

In other words, neither of us will be around when the loans default, so just go with it. Let's make some money.

'Finance has traditionally been an "eat what you kill" environment,' write Credit Suisse's Michael J. Mauboussin and Dan Callahan in a report describing short-termism in the financial industry.[36] 'The very nature of trading appeals to certain types of individuals who may be more focused on their own performance and making money – personalities that are often difficult to manage.'

Salaries at financial firms are often exorbitantly high compared with other professions, they continue, and this can foster more aggressive behaviour that favours risky, short-term wins over long-term prudence. 'It is seen in Western cultures as a measure of success and power to be paid handsomely, bringing with it a sense of bravado, while perpetuating a mindset of courageous decision-making,' say Mauboussin and Callahan.

The irony is that asset managers that have a stake in their own firm via shares, instead of being incentivised solely through salary and bonuses, saw greater gross returns over the longer term, according to a research study by FCLT Global.[37]

So, not only is there a culture of beating the gun in modern finance, but also an incentive structure that enables it, which attracts a certain kind of personality less inclined to the long view.

Another way that targets nudge people away from the long view is when goals are thought of as 'ceilings'. This is best demonstrated by the 'New York City taxi effect'. When it rains in NYC, it's hard to get a cab. Common sense would suggest that it's because demand is high, but it's also due to supply. Researchers have found that taxi drivers often don't choose to take full advantage of bad weather.[38] They could, of course, reap in fares all day. What actually happens instead is that they earn their day's target faster, and so clock off early. Rain encourages people to take a greater number of short journeys, which are more lucrative for taxi drivers, but after a while it leads to fewer cabs on the road.

So, while the rainy-day taxi driver could choose to, say, put away some savings for retirement, the temptation to put their feet up early is too strong. Few could blame them, it's a difficult job. But the point is that the taxi drivers would be better off in the long term if they ignored their daily target and worked their normal hours.

The historian Jerry Z. Muller, author of *The Tyranny of Metrics*, argues that 'metric fixation' underlies myriad modern problems – and it applies to many other types of organisation, not just in business. If police officers are promoted when major crimes go down, it may incentivise them to downgrade or simply not record crimes, he explains, and if surgeons have their success rate publicised, affecting their income and reputation, they may choose to improve their scores by refusing to operate on patients who need risky procedures.

'Almost inevitably, many people become adept at manipulating performance indicators through a variety of methods,' Muller writes, 'many of which are ultimately dysfunctional for their organisations.' This negative effect of metrics on people's behaviour is neatly described by a principle called 'Goodhart's law', named after British economist Charles Goodhart, which is often phrased as: 'When a measure becomes a target, it ceases to be a good measure.'

So, does this make capitalism inherently short-termist? Should we even seek to dismantle it altogether? Not necessarily. It has, after all, played a role in human culture for centuries, transforming living standards and lifting hundreds of millions out of poverty.

Take the long view, and we can see that there have been different forms of capitalism before, and so there could be different forms again. The Western-style version that is so dominant today only really took hold in the latter half of the twentieth century. After the Second World War, the Mont Pelerin Society, a think tank composed of academics, business leaders and economists, assembled with the aim of fostering free-market neoliberal values in the West. It was 1947, the year after Keynes died, but it's unlikely he would have been invited – nor would he have wanted to be[39] – since many of the Mont Pelerin members strongly disagreed with his ideas about the greater role of governments within capitalism. They instead wanted minimal intervention and regulation. These ideas eventually led to 'supply-side economics', an approach that truly took hold in the 1980s, with the Reaganite and Thatcherite doctrines of low taxes, small government and free trade.

While there have no doubt been upsides for many to this hands-off approach, it has also allowed time-blinkered habits to grow unchecked. And it is a version of capitalism that has shown itself to be markedly less equipped to respond to long-term changes, such as rising inequality, environmental change, or stagnation of household income, as well as failing to anticipate shock events like the 2008 financial crash or Covid-19 pandemic. 'In all these ways, therefore, the performance of Western capitalism in recent decades has been deeply problematic,' write the economists Mariana Mazzucato and Michael Jacobs.[40] 'The problem is that these failings are not temporary; they are structural.' However, that does not mean there are no solutions. 'Western capitalism is not irretrievably bound to fail; but it does need to be rethought.'

Mazzucato and Jacobs are far from the only serious economists or business executives to be making this point. As Dominic Barton, the former managing director of the consultancy firm McKinsey, has pointed out: 'The deficiencies of quarterly capitalism of the past few decades were not deficiencies in capitalism itself – just in that particular variant.' Capitalism needs renewal, he argued, not necessarily dismantling. 'By rebuilding capitalism for the long term, we can make it stronger, more resilient, more equitable, and better able to deliver the sustainable growth the world needs.'[41]

In recent years, various ideas and proposals have emerged for renewal. What they have in common is the idea that businesses need more varied targets for success than simply short-term profit and growth. In business, there's 'conscious capitalism', inspired by brands that trade on sustainability and environmentalism credentials. In policy, there's 'inclusive capitalism', advocated by both the Bank of England and the Vatican, which advocates harnessing 'capitalism for good'. And then there's the B-Corporation movement, which obliges certified companies to consider 'the impact of their decisions on their workers, customers, suppliers, community and the environment'.

Meanwhile, a number of individuals and companies have demonstrated that you can think longer – and succeed. As we learned earlier, Paul Polman of Unilever was initially punished by investors when he stopped quarterly reporting, but a decade later, Unilever was in much better shape, outperforming many other comparable companies. A number of the businesses Unilever now owns have thrived via their long-term credentials, including the environmentally friendly cleaning products of Seventh Generation or Ben & Jerry's, which is a certified B-Corporation.

When Polman eventually departed, the *Financial Times* described him as 'one of the most significant chief executives of his era', saying 'his approach to business and its role in society has been both valuable and path-breaking . . . he helped create a new discourse that speaks to those disillusioned with business and its doings'.[42]

McKinsey has analysed the short- and long-term habits of more than

600 US public companies over thirteen years, by looking at their investment plans, earnings and growth.[43] Over that period, they found that the
revenue of long-term firms grew on average 47 per cent more than the
revenue of other firms, with less volatility. Their market capitalisation also
grew $7 billion more than other firms, and returns to shareholders were
better. Finally, they added nearly 12,000 more jobs on average than other
firms. If all firms had done the same, the US economy would have added
more than five million jobs over the same period.

The approaches of Polman and others suggest that when companies
change the kind of information they disclose about the future, they
can attract shareholders more likely to favour the long term – so-called
'dedicated investors'.[44] Short-termist approaches, meanwhile, attract short-
termist speculators. 'Companies obtain the shareholder constituency that
they seek and deserve,' the influential investor Warren Buffett once
observed.

One individual who has become well known for this approach is
Amazon's Jeff Bezos, who – before he stepped down as CEO – would
regularly communicate the long-term principles of the company to the
market. In his first shareholder newsletter in 1997, he wrote that the
'fundamental measure of our success will be the shareholder value we
create over the long term'. He reprinted those words every year. 'As far
as investors go, our job is to be superclear about our approach, and then
investors get to self-select,' he explained.[45] (Bezos has also embraced a long
view outside business, with sometimes divisive results, via his investments
in space exploration and nuclear fusion. And he owns the land that hosts
the Long Now Foundation's 10,000 Year Clock, a symbolic project that
will tick for ten millennia, to which we'll return in Part III.)

Deploying this kind of long-term language and communication style
appears to matter. The researchers François Brochet, George Serafeim
and Maria Loumioti once analysed the transcripts of more than 70,000
earnings conference calls held by 3,613 firms. They counted words that
emphasise a short-term focus (such as 'next quarter' and 'the latter half
of this year', for example) and ones that point to a long-term view ('years'

and 'long run'). The managers who used short-termist language worked in firms that were more likely to cut R&D in years when they reported small earnings. And when the researchers looked at the shareholder composition, they found many more 'transient' than 'dedicated' investors.

'Managers should be aware that to a large degree, they are setting the tone,' wrote Brochet and colleagues. 'The language a company uses when talking to investors is a meaningful indicator of its orientation – and the investors listening in on calls that emphasise a short-term approach are a largely self-selecting group who like what they hear.'[46]

However, if Western free-market capitalism is to evolve into a version that is truly longer term, it may need to look for ideas outside the cultures that host it, towards nations who have not embraced its practices quite so wholeheartedly. One country that has managed to embrace long-term thinking in its businesses for centuries is Japan. Could it hold lessons for the rest of the world?

The 1,000-year company

Around a decade ago, the charismatic leader Masayoshi Son of the Japanese tech company SoftBank stood on stage and gave a vision of the long view unlike any I have ever seen from a major corporation in the US or Europe. It was bold in tone, a little eccentric in parts and simply fascinating.[47]

After a shareholder meeting, Son began with a sombre tone, talking about sorrow, loneliness and despair. One can imagine a few people in the audience shifted uncomfortably in their seats. He then moved on to happiness, self-realisation and accomplishment, arguing SoftBank's history made it well placed to launch an information revolution that would support human happiness. So far, so corporate.

But after a few more facts and figures, it got really bold. Son had titled his slides as a '30-year vision', but it turned out to be so much more. Instead, he said, he was going to talk about SoftBank's role in the next 300 years. 'Thirty years is not enough,' he said. What followed was a highly unusual series of slides that were often science-fictional. It had cheesy stock images, a quote from Mother Teresa, a comparison between the brain cells of a chimpanzee and a man, imagery of cloned sheep and discussions of humanity's extinction.

Son spoke about future technology centuries ahead, including brain-based computers, AI with emotions, cloning, telepathy, snake-like disaster robots, virtual reality, artificial organs and human–machine symbiosis. He forecast the human lifespan reaching 200 years by the year 2300. And he also briefly looked at the looming risks, including an 'unknown virus', a prescient foreshadowing of the 2020 pandemic.

'Look into the far distance whenever you get lost', read one slide. It also advocated the wisdom of looking 300 years into the past to see 300 years into the future, as well as emphasising enduring timeless traits of human nature, such as the need for connection and love.

Perhaps the point wasn't to make firm predictions – in a way, the details about future technologies were arbitrary. Instead, the effect was to show to shareholders (and the world) that SoftBank was thinking long term.

Since then, SoftBank has become an investment powerhouse, with a reach far beyond Japan. Through a $100-billion 'vision' fund, one of the world's biggest venture capital funds, the company invests in all sorts of long-term bets outside its core telecommunications business, from artificial intelligence to medical technology. It has acquired major stakes in companies like Uber, Alibaba, Sprint and T-Mobile. And while it has sometimes led to occasionally crushing losses – losing billions in WeWork, the office rental company – Son has persisted with his longer-term approach, describing his approach as a 'fleet strategy' that involves investing in a wide array of sectors.

While SoftBank is a relatively modern firm, founded in the 1980s, if it does thrive for 300 years, it would be in a tradition common within its home country. Japan is home to some of the oldest businesses in the world, with more than 33,000 businesses that are more than 100 years old.[48] These companies – from small family businesses like tea sellers to large construction firms – are known as 'shinise', which means 'old shop'. While there are a few European companies that are hundreds of years old – Italy's Beretta was founded in 1526, and Grolsch started brewing beer in 1615 – centuries-old businesses are far more commonly found in Japan.

Why have they lasted so long? There's no doubt that some reasons are culturally specific. For example, a number of Japanese corporations have a unique way to foster stewardship: the practice of adopting adult men into a family. By some estimates, more than 90 per cent of adoptions in Japan are adults, not children.[49] The new son takes the family name and swears allegiance to his new ancestors, eventually becoming Katoku, the head of the family. Even major firms like Panasonic or Suzuki have done this.

Researchers studying Japanese firms have found that companies with adopted heirs consistently outperform companies with blood heirs (which in turn outperform non-family firms).[50] Motivated by the honour of adoption, the practice encourages star managers to invest themselves in the long-term prospects of the company. Meanwhile, the prospect of displacement incentivises blood heirs to raise their game, mitigating the 'Carnegie conjecture', proposed by the US tycoon in 1899, that inherited wealth 'generally deadens the talent and energies'.

Culturally specific? Yes. Archaically gendered? Definitely. (In the West, boards with more women are associated with better long-term strategies, according to some studies.[51]) Nonetheless, this long-term incentive structure stands in stark contrast to the Western practice of managers milking the plant (see page 59) before moving on to new roles. What it shows is that, when individuals are given the incentive to invest in the long-term prospects of their employer, it can radically alter the fortunes and trajectory of a company's lifespan. Think back to the Timken company too – it was passed between five generations of the Timken family before it was broken up.

Corporate executive Arie de Geus argued that long-term companies have 'a sense of community ... where managers all consider themselves to be stewards of a long-standing enterprise'. When employees see themselves as part of a collective enterprise, it's easier to nurture the long view. An organisation run with short-termist principles is like a puddle, says de Geus, and so more vulnerable to extinction if the wrong conditions come along. By contrast, a long-termist organisation can be thought of as a river. 'Unlike a puddle, a river is a permanent feature of the landscape. Come rain, the river may swell. Come shine, it may shrink. But it takes a long and severe drought for a river to disappear.'[52]

But there are many other reasons that can explain the longer view of capitalism found in countries like Japan. One intriguing fact about shinise is that a large proportion of them provide services that never go out of fashion. Of the 1,000 companies more than 300 years old, 230 are in the alcohol business, 117 are hotels and 155 in the food industry.[53] But it's also a question of priorities, and the targets they set for themselves and their employees. Scaling up, maximising profits, reporting quarterly or increasing market are not top of the list. Instead, it's just as important, if not more so, to aspire to other goals, such as ethical and stable stewardship or passing the baton to the next generation. In general, Japanese firms also tend to be more risk averse, building up large cash reserves in case of crisis.

Traditions play an important role too. The president of Unsoudou, a company founded more than 130 years ago that produces woodblock prints and art books, told researchers studying shinise about the continual need to balance the legacy with innovation: 'I am grateful to my ancestors, because we are doing business with what they left us. I feel very strongly about this . . . If there is a business opportunity, I do not reject it. However, I do not think exploiting this opportunity is always the best option, because it is just a small part of a long history. What is important is to create a business that can live for a long time.'[54]

One illustration of this is Kongō Gumi, a construction company established in the year 578. The company refers to an idea called 'ie', which means 'house', and emphasises continuation. It also has specific guidelines that date to the eighteenth century:[55]

- Always use common sense.
- Don't drink too much, use obscene words, or har[...] towards others.
- Master reading and calculating with the abacus, and practi[...] craft] all the time.
- Give each task your full attention.
- Don't diversify. Concentrate on your core business.
- Be well mannered and humble, and respect status.
- Respect others and listen to what they have to say, but do not be overly influenced by their words.
- Treat employees with a warm heart and kind words. Make them feel comfortable and work with them heart-to-heart, but create an atmosphere that reinforces your role as the boss.
- Once you accept a job, do not fight with other people about it, especially clients.

It has persisted by meeting the enduring demand for shrines of the Shinto faith, surviving through various tumultuous periods – typhoons, wars and earthquakes. The leadership hasn't always made the right decisions – notably in the 1980s and 90s when an expansion into apartments, hospitals and hotels left it vulnerable to the capricious property market – but as of 2021, the company was still operating, deploying the same techniques its carpenters deployed 1,400 years ago.[56]

If some shinise have thrived by serving enduring needs, others have succeeded by continual adaption to societal change. NBK, a company founded in 1560 that started out making iron kettles, now manufactures high-tech machine parts. And then there's a firm you will have heard of that began as a maker of playing cards for the Japanese game hanafuda: the gaming giant Nintendo. Though these origins may feel a million miles away from the digital worlds of Super Mario and Zelda, Nintendo's core business is about play, a trait of human nature that will never go away.

In his analysis of the traits of long-term companies, de Geus pointed out that they have a specific kind of sensitivity to the world around them.

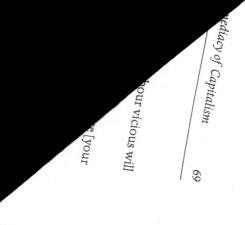

politics surge and ebb, they always

ut, staying attuned to whatever is

e in the longer term might be wise

o serve the timeless traits of human

ials and ores run out, demands and

how the human world evolves, there

desires that will always be there.

found within so-called *deep-time organ-*isations, which have _____ ss centuries. In 2019, the researchers Frederic Hanusch and Frank Biermann studied the commonalities between Kongō Gumi and several other long-lived organisations around the world, operating both inside and outside Western-style capitalism, including: Sweden's central bank Sveriges Riksbank (founded in 1668), Cambridge University Press (1534), the Marylebone Cricket Club (1787), the Royal National Lifeboat Institution (1824), Anti-Slavery International (1839), the University of Al Quaraouiyine in Morocco (859) and the hospital Hôtel-Dieu in Paris (829).

There were various common features. Many are 'benevolent monopolists', who have carved themselves a niche to provide services that speak to timeless human desires. They also tend to have close links with power and other long-lived institutions, such as royal families or religions.

But perhaps one of the most important lessons for the present-day company is that deep-time organisations are almost always recognised as delivering a broader societal change or stability that the wider community benefits from. They are often associated with a long-term public purpose that serves the common good – not just the shareholder or the customer.

Anti-Slavery International works 'to end slavery throughout the world'. Sveriges Riksbank promises stability through the doctrine 'hinc robur et securitas' ('herefore strength and safety'), which helps to guide its investing principles and guard against short-termist profit. And the University of Al Quaraouiyine provides an education underpinned by both the Islamic faith and benevolent concepts such as 'ummah' – Muslim community – social responsibility as expressed in 'zakat', and endowment via 'waqf', which advocates giving wealth to last beyond death. This is a principle that goes back to its founding in the year 859 when a Tunisian immigrant Fatima al-Fihri used her inheritance to fund a community mosque and school. It therefore predates Europe's oldest universities by more than a century (the University of Bologna was founded in 1088).

A typical company may not be able to follow every one of these lessons, but nonetheless there is much to be learned from the organisations that have been around a lot longer than the modern corporation. For long-term wisdom, Western-style capitalism instead often looks to its most recent successes, such as the inventors of Silicon Valley. Modern-day sages are often the technologists of the world, who talk of space exploration or hyperloop travel. Yet it is worth remembering that their companies are so much newer and more unproven than the deep-time organisations already embedded within societies.

According to the 'Lindy effect', a rule proposed by the writer and statistician Nassim Nicholas Taleb, if you want to know how long something will endure – an idea, a belief system or a culture, for instance – then ask a simple question: how long has it already existed?[57] Unless an entity has a fixed lifespan, the older it is, the more likely it is to survive into the deep future.

So, in sum we have found that many of capitalism's recent inventions and norms, from quarterly reporting to misaligned personal targets, have combined to become a key driver of time-blinkered behaviours in the twenty-first century. Just as there's no single source of blame, there's also no panacea.

We may have focused on capitalism in this chapter, but this can apply to any organisation. Bad incentives can be found in many realms of work and life, whether you work in healthcare, education, media, the charity sector or elsewhere.

Reassuringly, the experience of both long-sighted individuals and long-lived organisations show that it is not impossible to transcend the temporal stresses of the corporate short term. What's more, if leaders are prepared to delay gratification and communicate a longer-term vision, the evidence suggests that rewards will follow.

Capitalism's systemic short-termism is, however, not the full picture of how the time-blinkered age arrived. As we'll discover next, there are other major sectors of society that have combined to force a focus on the present – and it starts at the ballot box.

3

Political Pressures and
Democracy's Greatest Flaw

'It is a general principle of human nature that a man will be interested in whatever he possesses to the firmness or precariousness of the tenure by which he holds it.'

ALEXANDER HAMILTON[1]

'Governments must apply themselves to restore to men that love of the future with which religion and the state of society no longer inspire them.'

ALEXIS DE TOCQUEVILLE[2]

In the early 1980s, the US politician David Stockman gained political fame for his unrepentant short-termism and several indiscreet comments. Stockman was a young, ideological conservative. In the words of a journalist who wrote a notorious magazine profile at the time: 'He was thirty-four years old and looked younger. His shaggy hair was streaked with gray, and yet he seemed like a gawky collegian, with unstylish glasses and a prominent Adam's apple. In the corridors of the Capitol, where all ambitious staff aides scurried about in serious blue suits, Representative Stockman wore the same uniform, and was frequently mistaken for one of them.'[3]

Fresh-faced as he was, Stockman was also a canny political operator. He caught the attention of the President, Ronald Reagan, who in 1981 appointed him as director of Office of Management and Budget, one of the most powerful positions in the US government, overseeing where the money goes.

Reagan had been elected on promises of rooting out waste in federal spending, so Stockman knew it was his job to deliver. He embarked on a programme of huge, rapid cuts to the federal budget, slashing up to $40 billion. He had to move fast, with an eye on the next election. Soon there was blood on the carpet across Washington: the Department of Energy had its budget halved, a billion chopped from dairy subsidies, costs of Medicaid capped, eligibility tightened for food stamps and education aid slashed by a quarter. (Defence, though, had to go untouched, because Reagan had promised voters.)

'The classic young man in a hurry to realize his ambitions, he raced ahead, ramming through changes in programs, cutting here, slashing there, working, as he put it, "in a 20- or 25-day time frame",' wrote the *Washington Post* in 1981.[4] 'He became afflicted, like so many bright young operatives before him, with tunnel vision. Everything was immediate. The short term prevailed over the long.'

Soon though, Stockman encountered the kind of long-term problem that all politicians and their staff must eventually face. Pressure was growing to reform public pensions (social security), which was on a creeping trajectory towards bankruptcy. Rival Democrats proposed a plan to pump in money, but Stockman rejected it. His proposal, by contrast, involved drastic cuts to benefits for early retirees, on the grounds that it provided immediate relief.

When asked about his thinking, he notoriously said: 'I'm just not going to spend a lot of political capital on some other guy's problem in 2010.'[5]

It was a gloriously indiscreet thing to have said. And it would be quoted and requoted in academic books and papers about political short-termism decades later, as well as multiple newspaper articles and briefings about budget management.[6]

Stockman was nothing if not honest. There were – and still are – few incentives for someone in his position to make costly sacrifices that will only benefit future administrations. Sadly, honesty is not a valued trait in politics (another notorious quote of his that would come back to haunt him was: 'None of us really understands what's going on with all these numbers'). His transparency with journalists about his short-termism might have been tolerated, but combined with his candour about Reagan's economic policies, he eventually earned an unwanted invite to the White House.

There, he later recalled, the chief of staff Jim Baker told him: 'My friend. I want you to listen up good. Your ass is in a sling.' If it wasn't for me, Baker continued, you'd be fired. 'You're going to have lunch with the President. The menu is humble pie. You're going to eat every last motherf**king spoonful of it. You're going to be the most contrite son-ofabitch this world has ever seen.'[7]

Upon emerging from the lunch, Stockman said he had been taken 'to the woodshed' – a reference to punishment meted out as a child on his par-ents' farm in Michigan. He even suggested that he'd seen tears in Reagan's eyes. He kept his job for a few more years, but then moved to Wall Street.

Focusing on thankless, long-term tasks – on 'some other guy's problem in thirty years' – is a challenge for anyone in politics, no matter what their ideology and party affiliation is. As Barack Obama once ruefully reflected about his presidency: 'One of the hardest things in politics is getting a democracy to deal with something now where the payoff is long term or the price of inaction is decades away.' Or as Jean-Claude Juncker, the former president of the European Commission, once said about making tough but necessary economic decisions: 'We all know what to do, we just don't know how to get re-elected after we've done it.'[8]

The question is: could political short-termism be getting worse? The evidence is mixed. If you look across history, you might conclude blink-ered decision-making has always been there. In both democracies and autocracies since the beginning of civilisation, leaders have commonly acted in their own short-term self-interest, waged ill-advised wars rooted in rage or revenge, or over-exploited precious resources that led to their society's eventual collapse.

Along the way, there have also been rare flashes of remarkable fore-sight. A few visionary politicians have shown that it's possible to write long-lived constitutions, unlock new rights and privileges for women or minorities, or establish progressive institutions that endure for decades or even centuries. Others have built infrastructure that goes on to meet society's needs far into the future, from city sewers to transcontinental railway networks,[9] or established sovereign funds that distribute today's wealth to generations tomorrow.[10]

Longer-term plans have often followed crisis. After the Great Depression, President Roosevelt launched The New Deal to recharge the US economy, embedding social changes from state-supported retirement to the end of child labour. And after the Second World War, leaders internationally attempted to secure a more stable, progressive world for citizens with the establishment of the United Nations, the European Union and the UK National Health Service.

Often the political motivations have been economic, but not always. The desire to preserve what matters from the past and present has also occasionally benefited the future, with attempts to conserve heritage of historical importance, or the establishment of national parks that protect scenery and nature from damage and development.

It is clearly possible, then, for individual politicians to be pulled in either direction: to follow self-serving, short-term interests, or to take a longer view – and that's just as true now as it has always been. What I would argue, however, is that the influence of visible – and invisible – short-term temporal stresses has grown in the twenty-first century at the same time as more and more long-term problems have stacked up. And democratic politics has yet to catch up.

Democracy is one of humanity's great inventions but is far from perfect (or as the famous Churchill quote goes: 'Democracy is the worst form of government except for all those other forms that have been tried from time to time'). Despite the voice it gives to citizens, its incentives and deterrents are now creating particular dilemmas for the politician who wants to take the long view. Long-term thinking is a major challenge for policymakers, and yet with problems like climate change or pandemics, the stakes have never been higher. All politics is local, goes the adage – but as we'll discover, all politics is temporal too.

To understand why, let's first of all visit New England – where we will find broader lessons about the mismatched timespans of contemporary politics, and the challenges of balancing the management of a salient, urgent crisis with less visible longer-term problems.

Fast fires and slow burns

One recent winter, my family and I were living in Cambridge, Massachusetts, a short hop over the river from the city of Boston. Throughout the summer and autumn, friends and colleagues had warned to prepare for the snow, but when it came it was still a surprise. Back home in England, winter's palette has only greys. In Massachusetts, the flakes settle in white banks in a matter of hours, and even when it melts, blue-tinged icebergs cling to shaded spots for weeks. Every morning as I walked my daughter to the school bus stop, she would climb over the shovelled banks as I held her hand for balance, the crystals crunching beneath her boots.

What was a novelty for us Brits, though, can become an annual grind for the Bostonians. A single snow day can cost the Massachusetts economy $265 million.[11] In 2015, the region faced a particularly brutal winter, when more than 220cm (87in) fell in three weeks, shutting down the transport network, schools and business for days in a row.

Many roads initially went unploughed and removal was haphazard, and the blame fell on the city mayor, Marty Walsh. He made things worse by approving a parade for the New England Patriots through the streets, which angry residents felt diverted clearance resources from their own blocked-up neighbourhoods. At the same time, the Massachusetts governor Charlie Baker was drawing fire for the failing bus and rail network, immobilised by the weather. He got himself into an ugly political blame game with the Massachusetts Bay Transportation Authority.

Walsh and Baker must have known that snowstorms can doom a political career.[12] In 1959, New York City mayor John Lindsay was blamed for a crippling and deadly storm when ploughs failed to be deployed due to lax maintenance. What made things worse was that he and his limo got stuck in Queens, when he was supposed to be solving the gridlocked streets. His opponents pejoratively dubbed him a 'limousine liberal' – an insult that would endure for decades in the US political lexicon. It was a similar story a couple of decades later, when Chicago mayor Michael Bilandic allowed parking tickets to be placed on snow-buried cars. Or when Denver mayor Bill McNichols was blamed for a thirty-three-hour delay in street-clearance over the Christmas holidays.

Snow can be symbolic of the way that people judge how their city and state is run. When it falls in overwhelming volume, it is a trial of responsibility and legitimacy for a politician, and provides some of the clearest evidence a voter can see that their officials are capable of wielding the power entrusted to them.

When an event like this strikes – freak weather, an earthquake, a terrorist attack, a riot, or a bank run – the politician's role is clear. It involves action, task-forces, urgent meetings, press briefings and war-rooms. It is here that a local, short-term view is necessary. As the political scientist Simon Caney once pointed out: 'Focusing on a short-term goal is not necessarily bad or pernicious. Sometimes it may be permissible, or even required.'[13] I describe this necessary and conscious version of short-termism as *present-minded*: it's an important facet of the long view that we'll return to in later chapters.

Unfortunately, far too much of politics is shaped by the salience and sensationalism of the latest crisis. In the long run, many are hardly emergencies at all, whipped up by a competitive media or caused by politicians themselves: sex scandals, gossip or party-political squabbling. These issues deserve some attention, but not the disproportionate amount they so often get.

Meanwhile, there are problems evolving in the background, moving too slowly to draw attention. These changes move on the timescale of decades or even centuries – these are uncinematic and inconspicuous, but nonetheless can shape the world profoundly.[14]

We therefore might define two different types of political issue, characterised by their urgency, salience and pace: the ***slow burn***, and its opposite, a ***fast fire***.[15] Whereas a fast fire emerges and evolves quickly, drawing in all attention, a slow burn plays out in the background, often disregarded until it can no longer be ignored.

An obvious example? The slow burn of climate change. For Walsh, Baker and their successors in New England, a far slower, deeper problem than a winter storm looms. For the north-eastern US, climate change promises to bring extremes that will harm economies and wildlife for many years to come – and that includes heavier and more frequent snowstorms. A 2020 review in the journal *Nature* observed that, over the previous decade, disruptive snowstorms had hit cities on the US East Coast three times more often than the previous ten years.[16] One reason may be that accelerated warming in the Arctic is creating 'blocks' in the atmosphere, where weather systems slow down or fail to disperse as they move across the Earth's surface. This can affect the weather in lower latitudes. 'Like boulders blocking a river, once an atmospheric block forms, its impacts are felt both upstream and downstream,' the researchers write.

When such planetary-scale traffic jams happens, the winters in the mid-latitudes can get, well, weirder. Extreme, once-in-a-century events become more regular: potentially more catastrophic snowstorms, but also a greater frequency of disruptively warm periods, too.

I experienced a snapshot of this on one strange day in Boston in the middle of January. My family and I had been lucky to avoid the kind of freak storm that had hit Massachusetts a few years earlier, but what happened instead was unwelcome in a different way. On 12 January 2020, the temperatures in Boston were so high that they broke records, reaching 23°C (74°F). It's difficult to overstate how unusual this was. After weeks of down jackets and insulated boots, we spent a Sunday in T-shirts wandering Boston's North End, the city's little Italy, where cafés had hastily arranged tables and chairs on the street. My daughter's ice cream melted over her fingers. While the sunny respite was pleasant, such oddly balmy weather in January also made us feel uneasy. Especially so when the temperatures swung back below freezing only days later.

Weird weather is creating new challenges for politicians around the world. In New England, the strange seasons promise to bring heavier rainstorms with greater flooding risk, disruptions to fishing and farming, disease-carrying insects.[17] These effects are specific and local, so politicians are expected to tackle them, but the climatic causes transcend the edges of electoral sovereignty and accountability.

To their credit, some leaders have not been idle – Massachusetts has often been ahead of other US states on climate policy. Before he accepted a role in the Biden administration in 2021, Walsh led a group of more than 460 US city leaders called 'Climate Mayors'. But Governor Baker has had a more mixed history. Environmental groups once gave him a report card rating of 'A-' for climate resilience, but a mixture of Bs, Cs and an F on other issues.[18]

Environmental policymaking is far from the only area of politics where there are fast fires and slow burns playing out. As you can see below, a temporal lens can be applied across health, social, domestic and foreign policy too:[19]

ENVIRONMENTAL

Fast fire
    Extreme weather
      Natural disasters
        Environmental pollution
          Water supply
            Food supply / agriculture
          Resource scarcity
        Climate change
      Slow burn

HEALTH

Fast fire

Hospital scandal

Pandemic outbreak

Healthcare reform

Prevention of health problems, e.g. obesity

Pandemic preparation

Antibiotic resistance

Slow burn

SOCIAL

Fast fire

Violent crime

Immigration

Housing

Education

Pension reform

Intergenerational inequality

Slow burn

FOREIGN POLICY AND SECURITY

Fast fire

Terrorism

Espionage

Military deployments

Foreign aid

Defence investments

International treaties

Slow burn

ECONOMIC

Fast fire

Market crash

Interest rates/inflation

Energy prices

Unemployment

Tax policy

Financial regulation

Globalisation

Slow burn

TECHNOLOGICAL

Fast fire

Social media

Automation

Infrastructure

Gene editing

Artificial intelligence

Nuclear waste disposal

Slow burn

Not all of the political issues above are necessarily bad – some slow changes are positive; some disruptive events bring benefits. Nor am I suggesting that the slowest economic trends, for instance, move at the same pace as the slowest environmental or technological changes. But the point is that a temporal lens on the political agenda can help to reveal when too much time, energy and political capital is spent on the most attention-grabbing issues, which is usually at the faster end of the spectrum.

Importantly, slow burns and fast fires can also be entwined. Neglect the former and you get more of the latter down the line. Of all the political challenges we face in the twenty-first century, the most concerning is what could happen if we allow too many crises to pile up at once. The world experienced a snapshot of this in the Covid-19 pandemic, when governments lurched from one fast fire to another: PPE shortages, overrun hospitals, deaths in care homes, contact-tracing failures, economic fallout, vaccine supply issues and new strains. These were mixed in with salient political media stories: public figures who broke lockdown rules, the US presidential election, racial equality protests and more. I'd like to hope that politicians were simultaneously taking the long view during this period, but it seems unlikely.

The cultural historians Helge Jordheim and Einar Wigen of the University of Oslo argue that in the late twentieth century, an over-arching sense of Western 'progress' was replaced with a narrative of 'crisis', shaped by a concatenation of climate impacts, social upheaval, terrorism, migration, financial crashes and more. 'Rather than the international order being "in crisis", we argue that crisis is increasingly used to order the international,' the pair write.[20] This should serve as a stark warning: if all we can do is live in the present dealing with ever-more emergencies, blinkered to the dangers of tomorrow, it is a trajectory that leads to total collapse.

There's no doubt that avoiding such an escalation will be one of the most testing challenges of our generation. But what makes it even more difficult is that elected politicians are operating in democratic systems that were designed when there were fewer serious, global slow burns. Leaders in history certainly faced slow-moving or infrequent problems, but issues like climate change, biodiversity collapse or the threat of artificial intelligence were not on the agenda hundreds of years ago because technology, trade and progress had yet to create them.[21]

These global slow burns lead to other complications. From the beginning, democratic governance has relied on the concept of jurisdiction, and the application of power within that territory. Constituencies, states, nations. However, over recent decades, that principle has increasingly been tested. The modern politician must contend with issues that extend far beyond the borders of their electoral responsibility, both geographically and temporally.

The *geographical* challenge is well established. Due to globalisation, there are now myriad political issues that are fuelled by international trends, outside the scope of local decision-making. A factory with 2,000 jobs in Ohio or Glasgow, for instance, can be shut down by a boss in Beijing, yet it falls to the Midwestern or Scottish politician to deal with the fallout and blame.

At the same time, politics has become more and more *temporal*. Many of the grand challenges of the twenty-first century operate on timescales of decades or more. The trouble is that this timespan fails to line up with a political term. Whether you are a president, a governor or a local councillor, the length of your time in office – and the proximity of the next election – may prove impossible to ignore. There are typically 4–5 years between elections around the world, but in some places it's even shorter.[22] In Massachusetts, for example, the governor holds office for four years, but their counterparts in nearby New Hampshire and Vermont face elections every two years. So while there may be a short window after victory, campaigning begins not long after.

Some researchers have described this gap between term length and the span of political issues as 'time inconsistency'.[23] Others, like political scientist Simon Caney, call it 'the mismatch'.[24] Caney also points out that, even when a politician *does* decide to do the right thing, there's no guarantee their successor will carry the flame forward anyway, whether it's because the next administration holds other priorities or simply due to old-fashioned spite. He calls this 'intertemporal anarchy' to capture the idea that no government has any guarantee that its successors will pick up its recommendations or policies.

What all this means is that most of the incentives and agency lie in attending to the urgent, tractable fast fire, not preventing the slow burn. That will, in the words of David Stockman, be 'some other guy's problem' in thirty years' time.

*   *   *

Taking all this in, it might be tempting to conclude that the problem is democracy itself – and that other political systems would be better served to rule for the long term. Yet in authoritarian nations, present-day political priorities can be just as likely to foster time-blinkered decisions.

History is full of examples where undemocratic leaders clinging to power made decisions that harmed the long-term interests of their citizens. In the sixteenth and seventeenth centuries, for example, the authoritarian rulers of the Ottoman empire often made short-term decisions to kill or imprison all their potential heirs, benefiting themselves but reducing the pool of competent successors. In 1595, Mehmed III ordered his men to murder nineteen of his brothers and half-brothers by strangling them with a silk cord. Others locked up their relatives in ornate 'kafes' – or 'gilded cages' – hidden away inside their palaces.[25] Imprisoned as children, the captive heirs would often be comfortable and well fed, but they were utterly separated from the outside world. When they eventually ascended to the throne, this extreme isolation made them unfit to rule, with little knowledge of state affairs and poor physical and mental health (the rule of Ibrahim the Mad was particularly unhinged). The empire suffered in the long run.

In the present day, China is occasionally held up as an example of governance with a long view, but the reality is more complex. It would be easy to assume that, unencumbered by elections, it's easier for China to make long-term plans.[26] It certainly has the power to do things a democratic government cannot do unilaterally, such as displacing millions of people to make way for dam construction, or building entire cities to support industry.

In 2017, premier Xi Jinping outlined a long-term plan for 2035, which aimed to make China the world leader in innovation, with greater soft power internationally and a doubled GDP, followed by a second stage up to 2050 by which date the country will have become 'prosperous, strong, democratic, culturally advanced, harmonious and beautiful'.

Yet while Chinese leaders don't have to think about elections, they have continually faced other challenges that have nudged them towards present-day priorities and the conservation of their own power: events like Mao's death, the tumultuous events of 1989, the economic fallout of the Great Leap Forward, or the collapse of the Soviet Union.

Jinping may plot a strategy for the next fifteen to thirty years, but it could change. Broadly, China has actually had four successive 'grand strategies' since 1949, each of which has shaped the priorities of its politicians: revolution (1949–77), recovery (1978–89), building comprehensive national power (1990–2003) and rejuvenation (2004–present).[27] Each of these periods has also been marked by short-term self-interest and corruption, not to mention human rights abuses that continue up to the present day, such as the treatment of the Uighur minority.

The myth of authoritarian long-termism is borne out by an intriguing analysis of the long-term policies of governments internationally. The 'Intergenerational Solidarity Index', created by the researcher Jamie McQuilkin, uses nine measures to score nations on how well they are planning for the long term:[28]

Environmental

Forest degradation rate
Share of low-carbon energy consumption
Carbon footprint

Economic

Adjusted net savings
Current account balance
Wealth in equality

Social

Primary pupil–teacher ratio
Fertility rate
GDP-adjusted child mortality

In 2019, McQuilkin collaborated with the public philosopher Roman Krznaric – author of *The Good Ancestor* – to compare the performance of democracies with other political systems. While China's score placed it at twenty-fifth internationally, twenty-one of the countries higher in the rankings were democracies. And out of the twenty-five countries with the lowest scores, twenty-one of them were autocracies.[29]

Here's how the top ranks looks:

1.	Iceland	14.	Spain
2.	Sweden	15.	Sri Lanka
3.	Nepal	16.	Finland
4.	Switzerland	17.	Croatia
5.	Denmark	18.	Netherlands
6.	Hungary	19.	Bulgaria
7.	France	20.	Belarus
8.	Costa Rica	21.	Vietnam
9.	Belgium	22.	New Zealand
10.	Uruguay	23.	Italy
11.	Ireland	24.	Luxembourg
12.	Austria	25.	China
13.	Slovenia		

A triumph for Western democracy? Not quite. You may notice that the ranking does not feature many of the richest democracies either. Of the G7 countries, only France and Italy make the top twenty-five, whereas Germany is ranked twenty-eighth, Japan twenty-ninth, UK forty-fifth and Canada fifty-fifth.

And the US, that shining city on the hill? All the way down at sixty-second place – just lower than Russia (sixty) – pulled down the table by its high carbon footprint and surprisingly high child mortality rate compared with peer countries (driven by, among other things, expensive healthcare and less maternity support).[30]

There appears to be no single, obvious reason why a government scores highly, but there are some noteworthy correlations. For example, McQuilkin and Krznaric point out that politically stable states are more likely to score highly, which makes sense when you consider that a precarious, fragile government will always be more focused on short-term survival than intergenerational solidarity. Higher scores also correlate with decentralisation, which the pair speculate could explain why China (25) and Vietnam (21) perform better than the centralised governments of Egypt (103) or Saudi Arabia (98).

What's clear is that wealth does not boost a score. As well as the G7 nations, other countries with high GDP are similarly absent from the top ranking, such as South Korea at thirty-second, Australia fiftieth, Brazil fifty-fourth and India sixty-eighth. In some ways, this might be a cause for optimism. If smaller nations from Europe, Central America and Asia can top the rankings, it suggests that many other governments could in principle follow their example, starting with their longer-term social and environmental policies, for instance.

Unfortunately, if there are lessons to learned about long-term governance from these top-ranking nations, so far they have been ignored. And meanwhile, a number of the external forces that encourage short-termist politics may be getting worse. The question is: why?

Horizon creep and the Buxton Index

While blinkered behaviour has always been a problem within democratic politics, in the twenty-first century, there are a growing number of external pressures outside it that nudge leaders to a short-term view. In recent decades, politicians have been pulled in the direction of even shorter attentional timeframes than the electoral cycle. I call this *horizon creep*, and it can have a variety of causes.

Politics intersects with various other sectors of society with shorter views, and in many cases, their influence over decision-making has grown. Policymaking, after all, does not happen in a bubble.

A way to think about how these different sectors interact is by using a rough measure known as the *Buxton Index*. This allows you to see that politics has a different horizon to the sectors it intersects with, such as business, media, think tanks and so on.

Originally coined by John Buxton at Warwick University, the Buxton Index is defined as the average period, measured in years, over which an entity makes its plans.[31] A typical corporation might have a Buxton Index of, as most, 1–2 years, whereas a scientific advisory group focused on climate change might be more like 30–80 years.[32]

Across politics, the Buxton Index may vary. The departments for environment or pensions might have a longer horizon than the departments for business or culture. And a politician attending a United Nations meeting to discuss sustainability targets in 2030 might temporarily boost their index, only to pull it back down again with a David Stockman-esque decision on budget cuts. To be generous, a typical politician with an eye on the next election, but not totally dominated by their personal career prospects, might have a Buxton Index of around 5–10 years.

Plenty of the businesspeople who this politician encounters, however, may have a planning horizon that is shorter. Through lobbying and campaign funding, corporate interests have great sway on policy, nudging governments to focus on nearer-term economic growth, such as lowering taxes or loosening regulation. Political scientists have found that corporations have become the largest single interest group operating within political systems in the US and Europe.[33] Several decades ago, businesses were more likely to operate collectively via trade associations, lobbying on behalf of their shared concerns and the overall health and longevity of their sector. But individual corporations have increased their access to the political process on both sides of the Atlantic to become the dominant force, with each now lobbying on behalf of its own bottom line. (More broadly, capitalist pressures have also led to an economic practice called 'discounting', which devalues the well-being of future people and enables political short-termism to claim economic justification. It's complex, and its implications are far-reaching, so we'll return to it in more detail in Part III.)

However, capitalism is not the only reason for *horizon creep* within politics. There are also the pressures of the news media, which arguably has an even shorter view than business. To understand why this matters, let's look at how the relationship between journalists and politicians has changed in recent years.

Media matters

One day in 2019, while preparing to give a talk about the long view in journalism, I decided to look up the front page of BBC News exactly ten years earlier.

This was the list of top stories I saw, published on 24 February 2009:

> US recession may last until 2010
> Spain makes Guantanamo offer
> Nazi row bishop leaves Argentina
> Hudson pilot urges safety funding
> Obama hails US–Japanese alliance
> UK vetoes release of Iraq war dossier
> Failure hits Nasa's 'CO2 hunter'
> Italy and France pen nuclear deal
> Murdoch 'racist' cartoon apology

I was struck by the transience of the headlines on this page, and how little most of them mattered a decade afterward. With hindsight, how many had true longevity? Apart from the mention of recession, they didn't even tell me much about the most impactful event of that period: the aftermath of the global financial crisis, which had occurred only months beforehand.

Partly it's a question of habit: there are news media norms established last century that have yet to adapt to the slow-burn challenges of this one. For example, discussions about tax legislation in Washington DC tend to get a lot more column inches and airtime than, say, the threat of artificial intelligence, simply because issues like tax always have.[34]

But there's more to it than that. The decisions of governments are rarely covered with a long view in mind. Instead, the currency of political journalism is controversy: salient fights over issues in the present, with opposing actors, winners and losers. Controversy is visual too. If you're a TV producer, you can show politicians duking it out in parliament, press secretaries and presidents fielding aggressive questions at podiums, and broadcast vox pops with angry voters.

Try to do the same with long-term issues, and the journalistic palette is more grey: you have only sober two-way interviews and perhaps some stock footage to rely on. Reflective journalists will sometimes ruefully acknowledge that climate change should lead the news every day of the year, because it is the biggest story of the century. Yet it can't because it's not novel, and unless you can blame it for a wildfire or a hurricane that happened yesterday, it has no tangible imagery, because its most significant effects will happen in the future.

The news, by definition, is what happened on the previous day, or even less – and if it has any forward-planning horizon, it is usually measured in days and weeks, rather than years or decades. As Alan Rusbridger, the former editor of the *Guardian*, once put it: 'Journalism tends to be a rear-view mirror. We prefer to deal with what has happened, not what lies ahead. We favour what is exceptional and in full view over what is ordinary and hidden. There may be other extraordinary and significant things happening – but they may be occurring too slowly or invisibly for the impatient tick-tock of the newsroom or to snatch the attention of a harassed reader on the way to work.'[35]

These journalistic habits have existed since we started printing newspapers, but the power of the 'now' has amplified in recent decades. It began with the emergence of twenty-four-hour TV news in the late twentieth century, which entwined information and entertainment. A US campaigning group, called Rocky Mountain Media Watch, once captured the worst habits of the TV news industry by analysing the output of around

100 local news stations in the US.[36] Reading their findings as a journalist
myself is an uncomfortable experience – not least because what they
warned about my industry at the dawn of TV news has got far worse since.

They found three common themes dominating television news:

Mayhem: 'Stories about crime and natural and human-made disasters,
provoking excitement, fear, terror and alienation . . . such violence can
have a greater impact than fantasy or action entertainment, precisely
because it is real.'

Outliers, tragedies and shocks lead the news precisely because they are
unusual. In the aggregate, these stories can create an unrealistic picture of
the way the world really is. If something leads the news, it is by its very
nature likely to be rare or outside the norm. It also explains why political
journalism neglects the more boring, functional processes of governance.

Fluff: 'Comic relief to the darkness of mayhem: cuddly animals, cute
children, contests, and stories of the rich and famous. Fluff elicits a deep
emotional response in its viewers; it titillates and elicits warmth, humour,
familiarity, expectation and the excitement of voyeurism.'

There is of course, no reason why news can't be positive: but what
these stories often fail to capture is what the *actual* good news is: for
example, long-term declines in violent crime or global poverty, or greater
rights for once-marginalised people.

Pavlovian: In short, a 'predictable architecture'. News, due to its
presentation format, approaches stories within certain moulds. On TV,
information is funnelled through the treatment of talking heads. In politics, this might be the 'minister sex scandal', or framing a decision based
on updated information as a 'government U-turn'. Remove the individuals
and the issue, and the architecture of the story is always the same.

Since the Rocky Mountain campaigners looked at US TV before
partisan stations and social media really took hold, I would add an extra
theme that also now dominates the news and keeps attention locked in
the present: *Outrage.* The news today is no longer a passive experience,
only for digesting facts. Rather it serves as a fuel that encourages audi-
ences to express their identity, politics and beliefs online. Many cynical

media outlets know this, so serve up ever more via opinionated analysis and stories designed to trigger an audience response. On social media, outrage travels further.

The arrival of the digital age has supercharged these habits and norms. Fuelled by the internet, news cycles are short, loud and relentless. As you read these words, I'm confident that many of the most attention-grabbing, outrageous stories you read about online today will be hazily remembered a decade or more from now, becoming incremental footnotes within a broader historical story. Much of journalism now captures only a snapshot of the world, but by doing so, it pulls attention towards the transient present rather than identifying deeper patterns.

So, just as a politician turns their eyes to one problem, a new news cycle brings another, pulling them back into journalism's relentless timeframe.

It's perhaps not surprising that entwined with these developments in the media has been the rise of populism, and a breed of politician that thrives in this environment, adept at creating loud and transient news cycles to obscure their lack of long-term strategy, as well as polluting discourse with false claims. Former US President Donald Trump was notorious for manipulating the media in this way, moving the spotlight of attention away from his problems by tweeting or saying something distracting.

The practice is also captured by the 'dead cat' strategy, coined by Australian political strategist Lynton Crosby. As the former UK Prime Minister Boris Johnson once wrote of Crosby's approach: 'There is one thing that is absolutely certain about throwing a dead cat on the dining room table – and I don't mean that people will be outraged, alarmed or disgusted. That is true, but irrelevant. The key point, says my Australian friend, is that everyone will shout "Jeez, mate, there's a dead cat on the table!" In other words they will be talking about the dead cat – the thing you want them to talk about – and they will not be talking about the issue that has been causing you so much grief.'[37]

At this point, we could be forgiven for feeling despondent. With so much pressure on politicians to take the short-term view, it's difficult to see solutions. Yet the good news is that there is a growing number of individuals, organisations and activists trying to refocus governmental attention on the longer-term political view. In many ways, they have set themselves the most ambitious of tasks: not only are they asking politicians to orient their gaze away from present-day priorities, but they are also asking leaders to care about future generations who are not even born yet. Slowly, they are beginning to be heard.

The rights of future generations

Over the years, many have spoken on behalf of future generations. Thomas Jefferson argued each generation should be 'unencumbered by their predecessors' in their tenancy of the world. In 1866, John Stuart Mill gave a rousing speech to the UK House of Commons about the 'duty to posterity'. And Karl Marx once wrote that no society owns the Earth, 'they are simply its possessors, its beneficiaries, and have to bequeath it in an improved state to succeeding generations'.[38]

But one of the more unlikely political advocates in recent decades was not in politics at all. Instead he was known for the ocean – the French explorer Jacques Cousteau.

Cousteau gained fame as a scuba pioneer, conservationist and documentary presenter. But towards the end of his life, he also became a passionate believer in the idea that future generations should not be handed a damaged planet, so set out to change people's minds through political advocacy. The CEO of the Cousteau Society once recalled a conversation with the explorer about his approach: 'Forget the politicians – they all think short term,' Cousteau told him. 'But there is something I have found that works. Identify an issue with a campaign that has emotional appeal. Advocate a specific policy. Get letters, petitions, and faxes. With thousands of signatures, the politicians will join the parade – no – they will try to lead the parade . . . I think this is the only way.'[39]

In the early 1990s, Cousteau did just that. He wrote a 'bill of rights' for future generations. Drafted in clear, direct language, it called on politicians to acknowledge that each generation 'has a duty as trustee for future generations to prevent irreversible and irreparable harm to life on Earth and to human freedom and dignity'. Governments have a responsibility to ensure these rights, the document asserted. 'All appropriate measures . . . shall be taken to guarantee these rights and to ensure that they not be sacrificed for present expediencies and conveniences.'

He and the Cousteau Society then distributed a petition in support of the bill, which eventually racked up 9 million signatures from 106 countries – no mean feat in the pre-internet age. Later, the campaign inspired volunteers like Pierre Chastan, a former printer, to distribute the petition wherever he went.[40] Once while travelling by plane, Chastan managed to get the signatures of almost every passenger on board, and he would later deliver a barrel full of Cousteau's petitions to the United Nations in New York City, after sailing all the way across the Atlantic in a 10m-(33-foot)-long boat made from trees in his family's woodland, which he named *Message*. Wearing the same bright red beanie hat worn by Cousteau, he even managed to swing a meeting with the then-Secretary General Kofi Annan.

The Cousteau campaign for future generations managed to shine a light on an issue that, at the time, few in politics were seriously focused on. Sadly, Cousteau didn't live to see the full impact of his efforts, but eventually more policymakers did 'lead the parade', just as he had hoped.

The wheels of politics turn slowly, but they do turn. A few months after Cousteau died, the principles within his bill were incorporated by UNESCO into a declaration of its responsibilities towards future generations, which was approved by the UN General Conference.[41] And in 2013, the UN published a report called *Intergenerational Solidarity and the Needs of Future Generations*,[42] acknowledging the importance of the issue, and exploring ideas like introducing a High Commissioner for Future Generations. The concept is at the heart of the UN's attitude to sustainable development, which it defines as 'meeting the needs of the present without compromising the ability of future generations to meet their own needs'.

More recently, the UN made an even more significant commitment. In late 2021, the UN Secretary General António Guterres unveiled a report that weaves together Cousteau's concerns with more recent thinking on safeguarding the future. The report, named *Our Common Agenda*, outlined proposals to enfranchise tomorrow's generations and encourage international cooperation on the kind of crises that might endanger the very existence of the future.[43] Echoing Cousteau's language, Gutteres's report talked of forums to act on behalf of future generations 'as their trustees'. He suggested the appointment of a Special Envoy for Future Generations, who would advise the UN on how to represent and protect the interests of future people. And galvanised by Covid-19, Guterres proposed the creation of an 'Emergency Platform' of heads of state and other stakeholders that could be convened if a fast fire like a pandemic, nuclear event or biological attack emerges.

While a UN report is a declaration of will, not of certainty – requiring the support of member states – it nonetheless read as an exercise in long-term political thinking, emphasising the impact we will have on the long-term development of humanity without neglecting the present and its unique problems. It is significant that after thirty years, the UN Secretary General is finally leading the parade that Cousteau helped to start.

Meanwhile, a number of other groups and individuals around the world are also picking up the baton. Over the past few years, many others have also begun to converge on the idea that future generations should be represented and afforded rights in politics. There have been myriad efforts to integrate intergenerational rights into policymaking around the world – in Finland, Japan, Singapore, Hungary and Israel.

Intrigued by these efforts, I found myself watching a debate in the House of Lords in London in mid-2019. The subject was the rights of future generations, and an effort to encourage policymakers in Westminster to pay more attention to them when making their decisions.

The Houses of Parliament does not feel like a future-oriented place. Whenever I visit, I am struck by how much of the environment there is framed by the past compared with other legislative buildings around the world. The building itself is leaking and crumbling, and it is a place full of archaic practices and traditions. As we waited for the Lords debate to start, a shiny-shoed official frowned upon the men in our group for failing to wear ties.

Joining me in the Lords gallery to watch the debate was Roman Krznaric, who co-authored the analysis of the Intergenerational Solidarity Index we covered earlier (see page 86). He has been a prominent advocate for the rights of future generations in recent years – and in his book *The Good Ancestor* has called for a new movement made up of what he calls 'time rebels'. One of his arguments is that modern democracies, especially in the wealthiest countries, are 'colonising the future'.

'We treat the future like a distant colonial outpost devoid of people, where we can freely dump ecological degradation, technological risk, nuclear waste and public debt, and that we feel at liberty to plunder as we please,' he wrote in an essay for a BBC season I commissioned that year.[44] 'There is little that the unborn citizens of tomorrow can do about it. They cannot throw themselves in front of the King's horse like a Suffragette, block an Alabama bridge like a civil rights protestor, or go on a Salt March to defy their colonial oppressors like Mahatma Gandhi.'

A comparison can be drawn from history, he says. When Britain colonised Australia and displaced the indigenous people already living there, it drew on the legal doctrine now known as 'terra nullius' – 'nobody's land'. 'Today our attitude is one of "tempus nullius",' he writes. 'The future is an "empty time", an unclaimed territory that is similarly devoid of inhabitants. Like the distant realms of empire, it is ours for the taking.'

After milling around in the lobby, an official eventually ushered us to the gallery to watch the debate. We were there to watch Lord John Bird launch a future generations policy. His plan was to craft legislation that would compel policymakers to embrace long-term thinking in their decisions, later becoming his Wellbeing of Future Generations Bill.[45] If passed into law, it would compel public bodies to publish 'future generations impact assessments' alongside proposed policy decisions, while a specially appointed futures generations commissioner would review their performance.

Bird can be an abrasive man to meet in person, telling rude and inappropriate stories if you happen to be drinking a glass of wine with him in a parliamentary bar. His aides shift uncomfortably as he launches into each new anecdote. But there is no doubt he has done more in his life than many politicians over in the Commons to support society's most vulnerable and least represented. As the founder of the *Big Issue*, he has helped thousands of homeless people get back on their feet. This background and experience sets him apart from many of the hereditary peers he sits with in the Lords.

As Bird began the future generations debate, I looked down from the gallery at the other Lords, assembled on the red leather benches. The Speaker sat between them on the woolsack, a large cushion stuffed with wool, a tradition dating to the fourteenth century – originally as a reminder of how important the wool trade was at the time, but today better capturing how a country's biggest economic priorities tend not to endure in the long term. One Lord sitting directly below me had a cane perched against his bench. As the others spoke, he reached for his smartphone. For the next hour, he flicked his thumb across the screen, idly playing solitaire.

In his speech, Bird explained that he had woken up to the importance of future generations following his work with the homeless. 'I was once very much about the present, because the past had failed . . . the crisis was where I started,' he said. 'I'm sick of continuously mending broken clocks, I want to stop the clocks from breaking in the first place.' A majority of money and time, he continued, is spent on dealing with emergencies, like homelessness and addiction, instead of seeking to prevent the multigenerational traumas, failures and inequalities that lead to them. 'I don't think that we will find a way to do anything about poverty unless we reinvent the future,' he said.

Outside England, he explained, other governments have already begun these steps. Legislation to protect the well-being of future generations has already been passed into law close to home: in Wales. Following a new act in 2015, Welsh policymakers are now compelled to think about the intergenerational effects of their policies, and the country has appointed Sophie Howe to be its future generations commissioner.

I had met Howe on a rainy day in Wales, at an event I was organising for the BBC at the Hay Festival. A former senior leader in the police, she was passionate about her role and the Act underpinning it. 'It applies to all of our public services. Health boards, local authorities, national bodies, and really significantly the Welsh government themselves have to demonstrate how they are taking decisions which meet today's needs without compromising the ability of future generations to meet their needs,' Howe explained. While she doesn't have censure powers, she can at least name and shame short-termist politicians.

It might be fair to say that it has not gone totally smoothly – changing entrenched political bureaucracies takes time. And Howe's jurisdiction is restricted to a local or regional scale, her achievements including the cancellation of a road bypass that would exacerbate climate change, or lobbying for school exam reform. But the power of this Welsh experiment is that here is a real-world effort that has moved from theory to practice. Along with Finland and Sweden, which have introduced parliamentary advisory groups to foster a longer-term view, and Hungary, which also has an ombudsman for future generations, Wales has been inspiring other nations to reassess their own norms and cultures in governance.

Back in the Lords gallery, I watched as Bird continued to speak about the power of the Welsh example and how England could do the same. He didn't win universal agreement: Lord James Bethell, a hereditary Tory peer, expressed concerns. 'I worry about political interventions on behalf of people not present. I worry about people acting on behalf of a mother or father in the voting booth.'

But he was among the minority. Bird had support from a wide array of colleagues in both Houses. Before the 2019 UK general election, all of the main party leaders signed his 'future generations pledge' that asked MPs to promise to account for the rights of future people and to 'work to prevent problems, including the climate crisis and poverty, from happening, instead of dealing in short-term, political emergencies'. It might have only been lip service, but it was a start, and was notable for its cross-party support.

Howe's efforts in Wales and Bird's campaigning in England have served as a focal point for many others to assemble around, both inside and outside Westminster. And this future generations movement has spawned many other proposals for reform in recent years.

One of my favourite suggestions is a mischievous idea called 'an intergenerational chaining effect' proposed by the philosophers Tyler M. John and William MacAskill.[46] As an incentive to focus on the long view, the pair suggest that the generosity of a politician's pension should not be decided until the political generation that follows. It would provide a clear incentive for individuals to care about their long-term decisions. Woe betide the politician who hands their successors a shoddy legacy.

John and MacAskill also suggest that, eventually, a certain proportion of the legislature could be made up of elected representatives for the future. In the UK system, they suggest this might include an upper house in the legislative branch of government devoted exclusively to the well-being of future generations.

The political scientist Simon Caney has proposed other (perhaps more realistic) ways of redirecting political attention towards the long view. He suggests earmarking specific days in the parliamentary calendar as 'Visions for the Future Days', or a 'State of the Future Union' speech in which political parties, NGOs and others engage in public deliberation about the future. He also proposes that governments and opposition parties produce a 'Manifesto for the Future' ahead of elections.

Citizen assemblies or 'mini-publics' that represent future generations could be another route. Sometimes called 'Future Councils', this is a form of deliberative democracy that can even involve actually role-playing future people. This would build on the assemblies already being deployed worldwide to shape real-world policy, from Ireland to Belgium.[47]

I tried this kind of standing-in-the-shoes exercise briefly myself at a session in late 2020 organised by the School of International Futures, a non-profit foresight organisation. I was asked to imagine I was a man called Adam living in the 2040s, who ran an international supermarket chain, troubled by the attitudes of his two teenagers towards technology and their careers. Another person in my group role-played a government minister. In the role of Adam, I was surprised to find myself talking about priorities that are quite far-removed from my current life, such as Adam's relationship with his (non-existent) teenage children. It was only a short exercise, but the simple act of imagining the needs, views and priorities of a future person was an intriguing way to experience empathy across time.

This Future Council idea mirrors a practice that originated in Japan called Future Design. Led by the economist Tatsuyoshi Saijō of the Research Institute for Humanity and Nature in Kyoto, it involves asking people to wear ceremonial robes that place them in the shoes of future generations when weighing up the pros and cons of policies. This simple ritual has been shown to change how they think.[48] Saijō and colleagues found that people were more likely to approve a present-day tax that would improve long-term water infrastructure in their town in 2060 – and has led to tangible policy changes in places like the town of Yahaba. Saijō describes such acts of temporal empathy as 'futurability'. 'A person exhibits futurability when she or he experiences an increase in happiness as a result of deciding to, and taking action to, forgo current benefits to enrich future generations,' he writes.[49]

The path ahead

To break out of blinkered politics, there's no doubt that the scale of the task is huge. Liberal democracy is designed the way it is for good reason: short political terms may discourage a long view, for example, but they also protect against abuses of power. And for better and worse, the system may always be skewed in favour of voters living now.

Yet history shows it is possible for politics to change. Future people and their needs may be ignored now, but so were other under-represented groups in the past – women or Black Americans, for instance. Those who pushed for those changes decades and centuries ago would have felt similarly daunted as they embarked on their projects for a fairer politics.

In the 1830s, Alexis de Tocqueville wrote that the role of the politician was to shine a light on the future, not to obscure it. In a capitalist society, there is much to distract citizens, from fear of poverty to opportunities for fortune, he said, and 'in the midst of these "perpetual fluctuations of fate", the present grows large; it hides the future that is being effaced, and men want to think only of the next day . . . the range of man's sight is circumscribed'.[50]

Therefore politicians have a grander responsibility. 'Governments must apply themselves to restore to men that love of the future with which religion and the state of society no longer inspire them,' de Tocqueville concluded.

One thing I have found reassuring to observe over the years is that long-termism, in its broadest sense, can be a cause that all political sides can get behind. Short-termism is often associated with the most tribal, divisive politics. But when a duty to posterity is discussed, it tends to win bipartisan support. And when there is a call to pass on an undiminished planet to our children, it can join people in a shared, nobler mission than the day-to-day battles of party politics. Embracing the long view is not necessarily about making sacrifices. If we can learn to be more long-minded in our political systems, it could benefit us today as much as it does future generations.

Over the past three chapters, we have looked at the pressures that can nudge societies towards time-blinkered behaviours. We've seen how a variety of factors across culture, capitalism and politics have grown in influence, discouraging the long view: entrenched bad habits, misaligned targets and systemic stresses. Some are clear and visible; some are hiding in plain sight. But the good news is that it is not impossible to avoid those effects if we can identify what they are, and how they combine.

But to escape the time-blinkered age and take one step closer to a more long-minded society, we also need to gain a better understanding of the workings of the human mind. If Part I has been about the external pressures nudging people's behaviour, the next leg of our story is about looking within. Our goal is to trace how humans evolved the capacity to think about time in the first place and how people's psychology shapes their attitudes towards their past, present and future. The aim is to gain a much richer sense of what it takes to cultivate long-mindedness as an individual, avoid short-term biases and master our cognitive potential. And it begins with Virginia Woolf, an extraordinarily mischievous chimp and a man with one of the rarest brains on the planet.

PART II

A TEMPORAL STATE OF MIND
UNLOCKING OUR AWARENESS OF TIME

4

The Timekeeping Ape

'Time is invention or it is nothing at all.'

HENRI BERGSON[1]

The writer Virginia Woolf was fascinated by the perception of time. Influenced by the philosophers of the early twentieth century, she saw an apparent gap between the time of the clock and the time of the mind.[2] Whereas minutes and seconds ticked forward with predictable regularity, she saw the mental manipulation of past, present and future as considerably more fluid.

In *Mrs Dalloway*, the character Clarissa has free-flowing thoughts that wander between memories and plans – only to be interrupted by the regular chimes of Big Ben that drag her back to the present: 'Out it boomed. First a warning, musical; then the hour, irrevocable. The leaden circles dissolved in the air.'[3]

A few years later, Woolf expanded on this theme in a passage in *Orlando*:

Time, unfortunately, though it makes animals and vegetables bloom and fade with amazing punctuality, has no such simple effect upon the mind of man . . . An hour, once it lodges in the

queer element of the human spirit, may be stretched to fifty or a
hundred times its clock length; on the other hand, an hour may
be accurately represented by the timepiece of the mind by one
second. This extraordinary discrepancy between time on the clock
and time in the mind is less known than it should be and deserves
fuller investigation.[4]

So, taking a cue from Woolf, let's now embark on that fuller
investigation.

The scientific understanding of time perception has come a long way
since the philosophical theories of the early twentieth century. How the
mind wrestles with the past and future is far more complex than Woolf
and her contemporaries ever realised. We now know that the human
ability to think about events outside 'now' is far from a simple act. It's a
remarkable talent, and took millions of years to arrive on this planet. Life
had survived for billions of years without us having the ability to picture
the longer-term past or future. When it finally emerged in the human
brain, it was a new evolutionary invention.

Precious though this ability is, it is also not perfect. As we learned
in Part I, there is a range of cultural pressures that have converged
to discourage societies, businesses and political systems from making
longer-term decisions, nudging the individuals operating within them into
blinkered choices. But those external forces are not the only reasons why
taking the long view asks so much of us – to discover why, we have to
look deeper at the workings of our brains too. If we are to become more
long-minded, we need to acquire a clearer understanding of what's going
on inside our heads when the mind's eye transcends the present moment.

So, in the coming pages, we will explore what shapes our uniquely
human perspective on past, present and future, from its evolutionary
roots to people's psychological biases. What mental architecture have
we inherited from our ancestors? How well adapted is the brain to contemplating long-term timescales? And knowing all that we do about our
psychology, could we unlock new ways of thinking?

Our exploration begins by looking at how our abilities compare with some of our closest cousins in the animal kingdom – and in particular, a naughty chimpanzee who lives in Sweden.

One day, the keepers at Furuvik Zoo, a couple of hours north of Stockholm, noticed something unusual about the behaviour of one of their chimpanzees. A male chimp called Santino had begun methodically picking up stones and making little piles around his enclosure. His rock-collecting habit was not inspired by geological interest – but it did turn out to be an astonishing and apparently sophisticated case of mischief.

Early in the morning, Santino would calmly collect rocks from his moat. He also worked out how to break off fragments from concrete by knocking on the surface to see if it was fragile enough to remove. The piles he created were spread out in up to six separate places, under hay or behind logs, but almost all positioned close to the visitors' viewing area. At the time, he was the only male, and the female chimpanzees showed little interest in what he was up to.

When the zoo opened, the reason for Santino's ritual was clear. It usually happened just before lunch. As people gathered to gaze down at the enclosure, he wandered over to his stones, and began to hurl them at the crowd. The keepers rushed out to warn people about the projectiles, before ushering him inside, where presumably he received a thorough telling-off in Swedish.

Over the following days, the keepers tried to keep Santino away from the crowds, but this proved impossible because all they could do was attempt to command him with their voice. Instead, they had to stand guard to warn visitors not to get too close. One keeper described the events as Santino's 'hail storms'. After five days of the bombardment, a staff meeting was held, where it was agreed that Santino couldn't be outside any more during the day.

Santino's story represented more than a case of primate mischief. When scientists came to the zoo to study his behaviour, they were struck by the implications.[5] Here was tentative evidence of more sophisticated forward planning than had been thought possible in the animal kingdom. When the chimpanzee was calmly collecting rocks in the morning, was he imagining the joy of launching a hail storm of projectiles before lunch? On a basic level, it seemed so.

Yet Santino's behaviour was exceedingly rare. No other chimp had done the same, nor had similar behaviour been observed in the wild. In what was good news for zoo visitors but bad news for science, the keepers later decided that they had no choice but to do something that would dampen his aggression and stop his plotting, so we'll never know more about his unusual behaviour. Sadly for Santino, that intervention was castration.[6]

Is it a uniquely human talent to be able to move the mind's eye across past, present and future? For centuries, philosophers have thought so. Aristotle argued that other creatures are locked in the present.[7] And in the 1800s, Friedrich Nietzsche argued that animals are blessed not to carry the burdens of the past in memory, writing of cattle that 'do not know what is meant by yesterday or today . . . for whom every moment really dies, sinks back into night and fog and is extinguished forever'.[8]

Scientists are less certain. Santino's case isn't the only example that challenges our assumptions. Some captive bonobos and orangutans have demonstrated rudimentary forward planning in their tool usage during experiments.[9] Rats can apparently hold spatial memories of a maze to help them navigate it in future. And scrub jays have been observed preferentially hiding food in a place that they had learned would be empty or depleted the following day.

Still, the possibility that some animals could plan like human beings continues to be debated among evolutionary biologists.[10] What look like apparently sophisticated acts could be explained by other means. When decoding animal behaviour, still fresh in the minds of many researchers is the notorious case of Clever Hans the horse, who amazed people in the early twentieth century by apparently performing arithmetic and other intellectual feats. It turned out the horse was actually responding to involuntary cues in the body language of his trainer, and wasn't clever at all.

What scientists *do* know is that the vast majority of animals do not have the ability or potential to project their minds across long-term time like human beings can. While Santino and other animals might well be able to forward-plan over a span of a few hours, there is no evidence that they can flexibly project their minds across years, decades or centuries.

One of the main barriers may be that they lack complex language. 'One of the design features of language is *displacement*, the capacity to refer to the nonpresent,' writes the evolutionary psychologist Michael Corballis.[11] 'We are awash with fiction, gossip, stories around the campfire, the media – and for some of us, scientific publications. Historical account can take us back through centuries, and science (and science fiction) can also take us into imagined futures beyond the lifespan.'

It's not that animals are incapable of learning things that benefit their future selves. Your dog 'knows' who you are, your smell and your voice – and that you are a source of food and care. And as Ivan Pavlov showed, they will salivate in anticipation of a tasty meal. Even a sheep can be trained to recognise a face, as demonstrated by one delightful experiment where they were trained to successfully identify the faces of Jake Gyllenhaal and Emma Watson.[12]

But the dog or sheep have merely formed simple associations. While many creatures might appear to be acting to improve their welfare days and weeks ahead, their actions are closer to automatic. So, when the spider weaves its web, the bear hoards food and the salmon swims upstream, their behaviour is not informed by a recollection akin to a human memory, nor do they have a strategic plan for their future. Instead, their behaviour is an innate drive that emerged via natural selection millennia ago: they do what they do because that's what their successful ancestors did.

The size of the gap between humans and the vast majority of animals can be illustrated by the baffling mealtime habits of cebus monkeys. With dark haunches, light faces and curling tails, they are readily identified as the 'organ grinder' monkey, or the pet owned by Ross from the TV show *Friends*. Back in the 1970s, the scientist Michael D'Amato was watching a group of them in his laboratory, and was struck by their lack of foresight.

Every day at feeding time, the hungry monkeys in D'Amato's lab would voraciously eat biscuits until they were satiated. What, then, did they do with the surplus? Store the biscuits away for later? No, they'd throw the food at each other, and out of their cage, like frisbees. A matter of hours later, they were hungry again, and the behaviour would repeat.[13]

As Bertrand Russell put it in 1954: 'When an animal is hungry and food is before it, it eats on impulse, and there is not that gulf between the present and the future which is characteristic of conscious desire.'[14] Humans, by contrast, plan ahead to sate future hunger over days, months, or even longer. At our best, we are capable of thinking like Joseph in the biblical story, who knew that seven years of abundance would be followed by seven lean ones, and so encouraged the storage of surplus grain.

'Forethought is the most important of all the causes that make human life different from that of animals,' wrote Russell, and 'it has become gradually more dominant with the lapse of time.' From the advent of agriculture to the establishment of law, education and government, a long-minded approach to the future has underpinned our achievements throughout history.

Long ago, then, it appears we evolved the cognitive architecture to project our mental selves far across time. In fact, it may have been one of the key adaptations that enabled humanity to become Earth's most advanced species. How, when and why did it happen?

The origins of 'mental time travel'

Picture what life was like in the Pleistocene, a period that began around 2.8 million years ago. It was a time of particularly large mammals – mammoths and giant sloths – and dramatic climatic see-sawing between glaciation and back.

During this era, the genus *Homo* emerged. At some point, our pre-human ancestors moved from the forests to the savanna. It would have been quite the proving ground. In the open space, life became more dangerous and uncertain – hunting, foraging, migrating, fighting. The emergent hunter-gatherer lifestyle of these hominins would have involved frequent movement as resources depleted, and the need to plan for hunting: making basic stone tools, and travelling to killing sites. Later, long-distance migrations from Africa would have required ever-greater demands on memory and planning. *Homo sapiens* emerged around 300,000 years ago and by the end of the epoch, would reach every corner of the globe.[15]

There are many proposals for why our species was so successful in this era, from the harnessing of fire to our cooperative tendencies to the emergence of spoken language leading up to the 'cognitive revolution' of around 70,000 years ago. The flourishing of *Homo sapiens* is a story of myriad evolutionary adaptations, not to mention a good dose of luck.[16]

Yet none of the advances we made during our earliest prehistory would have been possible without the emergence of a more sophisticated awareness of the past, present and future. During this period, our ancient ancestors acquired the ability to time-travel – in their minds at least.

One of the first evolutionary psychologists to describe this step-change in human evolution – and why it mattered – was Thomas Suddendorf of the University of Queensland, Australia.

I first met Suddendorf in Sydney in 2016. He wore a floral shirt and a single black stud in his ear, and his German accent was untroubled by Antipodean influence, despite the decades he had lived there. Born in the small town of Vreden, 100km (62 miles) north of Düsseldorf, he had been drawn to science as a young man after tiring of the Catholic teachings of his surroundings. Instead, he resolved to explore the big questions about what makes us human, or as he puts it, why we became the peculiar species that we are.[17]

After a period of civil service driving an ambulance and then back-packing in South East Asia, he eventually ended up in New Zealand, working with evolutionary psychologist Michael Corballis at the University of Auckland. He lived on a houseboat in the swamps of Waiheke Island, powered by batteries and solar cells. It was there, in the 1990s, that he wrote a Master's thesis that would shape the rest of his life's work.[18] (Actually, he was forced to write it twice, because he lost the first draft after the power died in his houseboat.)

The specific puzzle that occupied him was the evolution of the human mind, and the cognitive gap between us and animals. What changed in human beings to allow us to shape the planet with such comprehensive dominance? Why did some of our hominin ancestors stay in the forests as we emerged to build cities, technologies and civilisation?

He proposed that a vital skill underpinning our evolutionary success was 'mental time travel', which allows us to recall both past episodes and construct future events. It means you can play and replay memories in the theatre of your mind – what you ate yesterday, your last holiday, your childhood home – and also mentally project yourself and others into possible futures: tonight's dinner, the beach you'll lounge on next year, the house you imagine retiring in.

Suddendorf argued that animals could not integrate their knowledge and behaviour into this rich canvas of the past and future in the same way as us. Mental time travel, he proposed, made us who we are.[19]

For our pre-human ancestors edging out on to the dangerous savanna, mental time travel would have offered a distinct and powerful advantage. Imagine two individuals about to enter a bear's cave: the one that hesitated outside, after imagining being eaten, would have survived longer than his friend who strode blindly in. After communication became possible, these early hominins would have realised they could share memories of food sources, for instance, or pass on warnings of predators or dangerous locations to be avoided in the future. Natural selection would have done the rest.

In the Pleistocene, mental time travel was not unique to *Homo sapiens*. Various hominins may have developed the skill. Archaeological evidence shows that *Homo erectus* made hand-axes, perhaps for butchering carcasses later. Their tool-crafting skill was apparently passed between generations, and they kept many more axes than they needed for everyday use – all in anticipation of the future. Later, Neanderthals and Denisovans almost certainly had the talent too. Evidence of Neanderthal burials suggests that they may have had the sense of a future beyond an individual's lifetime.[20] Yet as these ancestral cousins died away, the gap between the abilities of human beings and other species widened.

To understand how precious the skill of mental time travel has been for humanity – in the Pleistocene and ever since – you have to try to imagine what it would be like if it was taken away. It's a tall order to step outside our own consciousness, but we can get hints from the accounts of a few ultra-rare people. There are a handful of adult human beings who do not perceive time in the same way as you or me – they live without the ability to imagine the past or future – and psychologists have asked them what it's like.

The man with no tomorrow

Kent Cochrane was a man with a mind utterly different to other human beings. Born in 1951 in the suburbs of Toronto, he had a typical childhood and education. He was gregarious and something of a thrill-seeker. But aged thirty, he crashed his motorcycle, receiving a brain injury that left him with a specific and unusual form of amnesia.

If you asked him questions, he could recall plenty of facts: the capital of France, the definition of the word 'thermostat', or who Louis Armstrong was. He could also talk about time: how it works with seconds, minutes and hours, and what clocks and calendars do. When we retrieve knowledge like this, it's called 'semantic memory', and Cochrane's injury had left that aspect of his brain unaffected.

Instead, Cochrane's amnesia had stolen his recollection of past scenes and events. He had lost his 'episodic memory', the ability to recall or picture events in his past life. He failed to remember experiencing the death of his beloved younger brother a few years prior, or the time a potentially lethal chemical spill from a train derailment forced his family to evacuate their home.

Semantic and episodic memory are two separate ways of bringing the past to the present in the mind. Another way to understand this difference is to consider what you know of your life as a baby. You can name where and when you were born – that's semantic knowledge. But you have no memory of what your birth was like – that's episodic.

What's really interesting about episodic memory is that it also unlocks the future. Cochrane, who lacked it, was unique to science not only because he couldn't remember events that *did* happen, but also because he was incapable of imagining events that *could* happen. When scientists realised this, it was something of a surprise.

So, what was that like for Cochrane, to have no sense of the future? He was a calm, congenial and articulate man, who was able to describe his thinking. He reportedly took pleasure in answering the curious questions of psychologists, despite not remembering meeting them before. He would speak softly, often with a broad smile, seemingly discovering his own words at the moment he said them.

Here's a brief exchange from one of the interviews he gave in the 1980s with the psychologist Endel Tulving, who spent years studying him:[21]

Tulving: 'What will you be doing tomorrow?'

[15-second pause]

Cochrane: [smiles faintly] 'I don't know.'

Tulving: 'Do you remember the question?'

Cochrane: 'About what I'll be doing tomorrow?'

Tulving: 'Yes. How would you describe your state of mind when you think about it?'

[5-second pause]

Cochrane: 'Blank, I guess.'

At other times, Cochrane was asked to describe this 'blankness'. He said it is 'like being asleep' or 'like being in a room with nothing there and having a guy tell you to go find a chair, and there's nothing there'. On another occasion he said, 'It's like swimming in the middle of a lake. There's nothing there to hold you up or do anything with.'

Cochrane's case provided some of the first empirical evidence that we need to picture episodes of the past in the theatre of our minds in order to project ahead into the future.

This makes sense if you think about it. If I ask you to imagine what you will do after you put down this book, you can form a scene in your head. Perhaps you'll drink a coffee, step off a train, or lie down in bed. Crucially, your brain can only create those pictures by stitching together episodic memories of these acts into a tapestry of tomorrow.

In some ways, we are all born like Cochrane. Very young children do not remember events in the same way as adults, and they show little awareness of the future. While they learn voraciously from birth – whether it is to recognise the sound of their mother's voice, to gabble new words, or toddle on two feet – they show few signs of episodic memory or forethought for a few years.

Thomas Suddendorf and his colleagues have shown that children only acquire the skill of mental time travel around the age of four, successfully completing experimental tasks that test their ability to think ahead and satisfy their future needs.[22] These experiments were different to the famous 'marshmallow test', which gauges the ability to resist gratification (we'll return to that later). Instead, the test was inspired by an Estonian children's story about a little girl who wants to eat a dessert. In the tale, the girl dreams about going to a birthday party where the guests eat chocolate pudding, but sadly she cannot because she has no spoon – everybody must bring their own. So the next evening, the girl goes to bed with a spoon in her hand and hides it under her pillow.[23]

Suddendorf and colleagues decided not to conduct an ethically problematic experiment with a spoon and a chocolate dessert, because it risked making pre-schoolers burst into tears. So instead, they showed a series of three-, four- and five-year-olds an empty *Bananas in Pyjamas* puzzle which was missing its pieces. Later, in a separate room, those children were given the opportunity to choose a toy to take back to Banana-Pyjama room. It could be the missing puzzle pieces, but also one of three other toys. The four- and five-year-olds chose the pieces so that they could satisfactorily complete it, but only half of the three-year-olds did.

On the day his four-year-old son 'passed' one of his tests, Suddendorf writes about a surprising moment that happened later on.[24] The boy put a hand on his father's leg and said: 'Papa, I don't want you to die.'

Suddendorf gulped. 'Neither do I.'

The four-year-old continued: 'When I get bigger, I will have children, and you will be a grandpa, and then you will die.'

It was as if his son's mind had opened up, Suddendorf recalls, suddenly pondering the existential questions posed by his newfound mental time travel.

If our ancient ancestors in the Pleistocene had never evolved this skill – if they had retained the thinking of a very young child, or even developed something more complex like Cochrane's mind – perhaps things would have turned out very differently for *Homo sapiens*. Perhaps our species would have felt more content, untroubled by the contemplation of mortality or darker futures. There is, after all, a sense of security and stability that comes from being unaware of time's trajectory, even if it is unfounded.

Cochrane, incidentally, described himself as happy, and he lived a healthy life until his death in 2014. In an early interview, Tulving asked about his well-being:

Tulving: 'How would you say your life is in general?'

Cochrane: [smiles broadly] 'I don't know, it's liveable I guess.'

Tulving: 'On a five-point scale . . . where would you rate it?'

Cochrane: 'Four.'

Tulving: 'Do you have any complaints about anything?'

Cochrane: 'I don't think so.'

Tulving: 'How do you feel about yourself?'

[thinks for 8 seconds]

Cochrane: 'Pretty good, I guess.'

However, without mental time travel, there's no way we could have built modern civilisations, invented life-saving medicines, created great works of art, visited the Moon, or discovered our place in the Universe. It is, as Suddendorf's collaborator Michael Corballis once put it, the 'essence of imagination'.[25] When the skill combined with our species' urge to connect our minds and our talent for language, it set us on an astonishing trajectory from simpler life to where we are today.

The psychologist Martin Seligman once wrote that we might have chosen to call ourselves *Homo sapiens* – the wise human – but that's more of a boast than a description, and we would be more accurately named *Homo prospectus*.[26] 'We thrive by considering our prospects,' he argued. 'The power of prospection is what makes us wise.'

But that's far from the last word in terms of how our minds tackle time. What we have yet to approach is our *psychology*.

Imagine I offer you a choice: you can take $100 today, or $120 in a year. What would you decide? Posed with a question like this, most people elect to take the lesser sum now, rather than delay gratification for a bigger reward later. When people picture their future selves, the person they tend to see is something of a stranger.

This 'present bias' has been observed across many realms of life, from people's inability to save for retirement to their willingness to binge on sugary, fatty foods. It explains why people fail to shell out for home insurance, why they experience buyer's remorse after an impulsive purchase, or why a child will eat a single marshmallow in front of them rather than wait patiently for two.

But this well-known inability to delay gratification only scratches the surface of our temporal psychology. As we've learned, humanity's perspective of the past, present and future may be unique in nature – but to achieve that ability, our species has had to apply a few shortcuts along the way. And those mental tricks are imperfect, carrying invisible habits and biases that subtly influence our attitudes and behaviours. As the influential psychologist Daniel Kahneman once wrote: 'The mind is good with stories but does not appear to be well-designed for the processing of time.'[27]

So how do these psychological blind spots shape people's thinking?

Are some more prone to them than others? And how might they be avoided? For answers, let's start by looking at the hidden meaning within a famous painting about Icarus, which, as we'll discover, has more layers than first appears.

5

The Psychology of Yesterday,
Today and Tomorrow

*'Sympathy with persons, remote from us [is] much fainter than
that with persons, near and contiguous.'*

DAVID HUME[1]

Landscape with the Fall of Icarus, *with detail overleaf.*

The first time you see the painting *Landscape with the Fall of Icarus*, composed in the 1500s, it can take a moment to spot where the actual disaster is happening.[2]

In the foreground, a farmer is depicted with a plough and horse in rich detail. You can see the individual leaves of bushes and the folds in the man's robes. As your eye moves around the painting, you might look at the mountains on the horizon, or the ship in the ocean, its sails captured by the wind. It is perhaps only then that you notice Icarus, in the lower right corner, his legs kicking above the waves after falling from the sky. You can just about make out a few feathers drifting down to the ocean's surface, but not the wax melted by the Sun. He is drowning, while life elsewhere carries on.

W. H. Auden referenced the scene in his 1938 poem 'Musée des Beaux Arts', observing that the artist had sharply captured a universal truth: the leisurely indifference of bystanders despite human suffering in the background.

But the painting contains even deeper meanings – it can also serve as an introduction to how the mind builds a landscape of the near and far future.

To understand why, first we must step into the farmer's perspective, which is shared by the artist and observer. It would be easy to label his attitude as callous and uncaring, but the psychology is more complex than that. The farmer's response can be partly understood by how the mind processes events that are near or far. The distance from Icarus's drowning is not just geographical, there is also *psychological distance*. If, through proximity, he was more aware of the grim reality – if he could hear Icarus gasping for air or pleading for help – perhaps he would act, but the details of his death are only roughly known to the farmer, and therefore more abstract.

This, sadly, is a universal human truth. As the Stoics first articulated with their concept of 'oikeiōosis' – roughly translated as 'affinity' – a person's proximity affects our sense of moral obligation and empathy towards them. Others would later imagine it as a series of concentric circles, where family and friends occupy the innermost ring, and the stranger in a far-off place languishes in the outermost.[3]

So, what's going on in the mind? In the late 2000s, the psychologists Yaacov Trope and Nira Liberman decided to find out, exploring how and why people experience psychological distancing, and what causes it. What they came up with is a framework called 'construal level theory'.[4] It would become a highly influential idea in psychology and beyond, because it underpins so many aspects of how we navigate the world, on a personal level and at a societal scale.

In short, Trope and Lieberman's theory describes 'the subjective experience that something is close or far away from the self, here, now' – and all the ways that this shapes our decisions and behaviours. It is crucial for understanding how we perceive events in the past and future too. Why? Like the farmer who is separated from Icarus across *space* – we can become disassociated across *time* too. The further away an event is chronologically, the more its details and contours remain unseen.

Near and far effects

When you think about time, the mind does a simple but remarkable thing: it converts it into physical dimensions. If you imagine the past and future, you can't help but place the self within a temporal landscape. It's so deep-rooted and familiar, you may not even realise you're doing it. So, in English and many other languages, the past is 'behind' and the future is 'ahead'. Sometimes the mental self is traversing that territory ('Phew, we got through it. Now let's look forward to the weekend'), while at other times we are stationary, and events are moving towards or away from us ('Winter is coming' or 'The days and nights have passed').

What it means is that people conceive the deeper past and future as geographically further away from their experience in the present compared with events that are 'nearer' in time.

This unconscious association of *far* time and *far* distance crops up in many stories and popular culture. Children's fairy tales about the past begin with the words: 'Once upon a time, in a faraway kingdom'. Similarly, the opening line of *Stars Wars* is: 'A long time ago, in a galaxy far, far away . . .' (If George Lucas's movies had begun 'A long time ago, very close by', it would have made sense, but might have felt less exotically alien.)

Our stories about the future may well be influenced by the effect too. The economist Robin Hanson once made the speculative but intriguing suggestion that it explains why the future is often depicted in bluer tones in science fiction or corporate technology marketing. There are exceptions, but in movies with spaceships or futuristic cities, cooler colour grading and bluer, cleaner set designs are often chosen over the more realistic multicolour palette of the present day. The reason, Hanson argued, is because of how the eye perceives light across distance.[5] When you stand on top of a mountain and look out across the countryside, the more distant hills will seem bluer than the ones nearby, since blue light scatters more readily in the atmosphere than red light. Painters have long known this colour trick too. In *Landscape with the Fall of Icarus*, the mountains in the distance are painted with a cooler palette, while the farmer's field is reddy-brown.

So, the further you mentally time-travel into tomorrow, the further away it can feel in terms of physical distance too. When the future is understood as a landscape, it means tomorrow is *near*, but next year is *far*. Next decade or century is very far. And a millennium from now? Well, that's so far beyond the horizon, it's not even on the landscape. The deep future is a foreign place, a sketchily imagined land across an ocean.

And this brings us back to the predictions of Trope and Liberman's construal level theory. It can be helpful to think about it by considering two different lenses: the **'near' lens** and the **'far' lens**. When the mind visualises something as near or far, it has many unwitting associations with other attitudes and emotions in relation to the self, as described in the table below.[6]

Near	Far
Here, now, me, us	There, then, them, others
Concrete, detailed	Abstract, schematic
Cluttered and complex	Simple and roughly drawn
Thought of in 'How' terms (e.g. how will I get there?)	Thought of in 'Why' terms (e.g. why do I want to go?)
Important	Unimportant
Feasible	Desirable
Realistic	Optimistic
Pictures	Words

So, as well as making us care less about events and people distanced from us in the present (and our own future self too), this suggests that the perspective we have of the 'far' future is also not particularly realistic either. Construal level theory predicts that when the future draws near, we are more likely to think about it in a concrete, pragmatic and tangible way, but when the mind moves ahead across months and years, then the picture becomes ever-more abstract and roughly drawn, and crucially, the related emotions are different too.

You can see a more everyday example of the 'near' and 'far' lenses in how people tend to look ahead to their holidays. Imagine you have just booked a trip to a sunny destination next year. When you are thinking about it now, it's common to focus on the optimistic desires and the 'why' of the trip: the opportunity to kick up your feet on the beach, relax and sip a tropical drink. It's a broadly drawn picture that, experience inevitably shows, usually doesn't reflect the full reality of what holidays are like. What you tend not to imagine until the experience is more imminent are the concrete 'how' details of booking a taxi to the airport, queuing in passport control, or finding the hotel after you land. And you'll also screen out the possibility of negative stuff: the bad meal, argument with your partner and rude waiter. You'll get to sip that tropical drink too, but the further away your vantage point is in time, the less complete the picture.

In short, the further the mind travels into the future, the more that abstract – and often optimistic – desires and choices tend to dominate, but as the present approaches, these are overridden by pragmatic and tangible realities.

It is, of course, necessary for the brain to paint the further future with a more simplified brush: generalisations help us plan ahead and approach uncertainty. When the details of the future are unknown, the brain acts a bit like an astrologer. In a horoscope, astrologers make forecasts such as 'you will face an important decision today' or 'powerful issues will arise in your life that will make it difficult to find peace'. These are often not inaccurate, but they are also rarely *precise*. As Trope and Liberman point out, it would be a soon-to-be-unemployed astrologer who predicted that 'next Tuesday you will sit on the train next to a 52-year-old Latvian orthodontist; he will sneeze in your direction, which will result in you catching the flu virus'. To navigate the world effectively, the mind plays this trick too: starting rough, but then adding ever-more precise details as an event approaches.

But it's important to be aware that, when the mind does this, it also influences our moral attitudes. Long before construal level theory emerged, the eighteenth-century philosopher David Hume observed how his ethical priorities changed as he moved his mental perspective across near and far time. 'In reflecting on any action, which I am to perform a twelve-month hence, I always resolve to prefer the greater good, whether at that time it will be more contiguous or remote,' he wrote. 'But on my nearer approach, those circumstances, which I at first over-look'd, begin to appear, and have an influence on my conduct and affections. A new inclination to the present good springs up, and makes it difficult for me to adhere inflexibly to my first purpose and resolution.'[7]

When you apply the lens of 'near' and 'far' to some of the twenty-first century's long-term challenges, such as climate change, it becomes easier to see why it can be difficult for people to feel ethically motivated to act. When people face problems in the near present – unemployment or inequality, for example – they are far more concrete than a future problem that feels far. If these present-day hardships bring distress, perhaps prioritising them is the natural and humane way of being. But that doesn't mean that potential suffering in the future should be ignored either, for oneself and for others. As the philosopher Derek Parfit observed in the 1980s: 'When we imagine pains in the further future, we imagine them less vividly, or believe confusedly that they will somehow be less real, or less painful.'

Construal level theory predicts that people imagine a future world degraded by the harmful acts of the present as a wholly different place, not where they live now. But of course, that's wrong: when our great-grandchildren are born, they won't be far at all. Unless you jump ahead thousands of years, they will be on the same planet. Chances are, they'll live in the same country or city. They might even call the same neighbourhood their home. And like Icarus in the painting, they will have the capacity to experience the same suffering as any human being alive today.

So how might near–far effects be avoided? As with all cognitive biases, it's a little easier to point it out in others than diagnose oneself. But simply knowing what the mind is doing to build that temporal landscape can be helpful. And it may close the psychological distance if you can deliberately seek to paint the future with more tangible, vivid details.

The distant future is more of a challenge, but some psychologists have shown that it is at least possible to bring people's 'present' and 'future' selves closer together, even though they might be separated by a human lifetime. In experiments, people shown images of elderly versions of themselves were subsequently more likely to allocate funds to a retirement pot.[8]

Other research has applied a technique called 'perspective taking' which involves stepping into the shoes of a specific person in the future affected by negative environmental changes in order to foster greater empathy and action today.[9] People were presented with a story and images about a specific woman living in 2105, who explained how her life had been made worse by environmental changes. For example, she talked about how she burned her hands after going outside without sunscreen, or how she'd got a bad skin rash from bathing in a polluted sea.

Compared with those who hadn't heard the woman's story, or those who had been asked to be more 'fact-focused', the people asked to focus on taking the woman's perspective were more likely to engage with the issue of climate change afterward, as measured by a questionnaire, spending time reading about it, or taking away a brochure to read.

The perspective-taking effect would also explain the power of the Japanese 'ceremonial robe' research that we encountered in our exploration of politics in Chapter 3. Recall that encouraging people to wear the clothes of an imagined future person, metaphorically stepping into their shoes, made their needs and desires feel more salient, fostering temporal empathy and greater support for political policies that would benefit their descendants.[10]

Artists, too, have experimented with perspective taking. For example, the London-based design studio Superflux – led by Anab Jain and Jon Ardern – once encouraged people at an exhibition in the United Arab Emirates to take a gulp of polluted air from the future. They created a series of air samples from the years 2020, 2028 and 2034, containing the most likely combination of pollutants, based on climate and fossil fuel emission projections. 'It was noxious stuff, impossible to inhale even a small amount,' the Superflux designers later recalled. 'Such experiential evidence brought home the point that often predictions and data can't.'[11]

All this suggests that if we can find techniques to deliberately and consciously paint the future more *concretely* in the landscape of the mind, it changes other, more emotional, attitudes attached to the near and far lens. People in the future (including oneself) become less like strangers – like Icarus in the distance – and closer to 'here, now, me, us'.

In Part III, we'll explore this in more detail – looking at the ways that stories and symbols can help make long-term time more accessible and psychologically closer. But first, let's continue to explore the mind's temporal habits and biases, because the near–far effect of construal level theory is not the only psychological lens to influence people's temporal decision-making.

Next, let's look at a little-known event following the first Moon landing in 1969. Shortly after Neil Armstrong, Buzz Aldrin and Michael Collins landed back on Earth, their bosses at NASA made a very unwise decision. It was a rare time in history when a convergence of psychological biases could have ended the world.

When rockets and probes venture into the Solar System, they get a thorough clean before lift-off. It's to avoid accidentally sending microbes into the cosmos. If extraterrestrial life was eventually detected, we'd want to be sure it wasn't an Earth-based organism that hitchhiked in the crevice of a dirty spacecraft. Worse, accidentally seeding life on other planets could kill all the creatures that were already there.

However, you can't easily clean a lander or vessel once it has left Earth, so there is another danger that's perhaps lesser known: if it returns home, it could carry hazardous alien life on board. Not little green men, but some form of microbe that could, in principle, catastrophically outcompete life on our planet or consume all our oxygen.

When NASA was planning the first Moon landings, the probability of picking up aliens was not considered likely, but many felt the risk had to be explored, because the consequences were so severe. 'Maybe it's sure to 99 per cent that Apollo 11 will not bring back lunar organisms,' wrote the scientist Carl Sagan at the time, 'but even that 1 per cent of uncertainty is too large to be complacent about.'

To assuage these concerns, NASA reluctantly agreed to install an expensive quarantine facility on the ship that would pick up Armstrong, Aldrin and Collins. The protocol stated that, after the astronauts splashed down in the Pacific, they should stay sealed inside the spacecraft, and a crane on the recovery ship would lift the whole module on board. The astronauts would then spend three weeks in isolation before they could hug their partners or shake the hand of the President.

They were indeed quarantined, but when the day came, there was a major gap in the procedure. After splashdown, the astronauts waited inside the capsule, bobbing up and down on the ocean surface. It was hot, uncomfortable, and they hadn't showered for more than a week. So, at the last minute, NASA officials decided to make things more pleasant for their three national heroes. They told them they could open the door.

In principle, this moment could have been a pivotal turning point for humanity. As soon as the capsule was opened, all the air inside flooded out. If it had been carrying lunar life, that decision to prioritise the short-term comfort of the men could have released deadly aliens into the ocean.

If this worst-case scenario had happened, it would have been an event that the law scholar Jonathan B. Wiener of Duke University has called a 'tragedy of the uncommons' – an event that happens extremely rarely but can be catastrophic. Whereas its better-known sibling effect, the tragedy of the commons, describes how selfishness damages a resource that benefits everyone – such as a forest or a water source – the tragedy of the uncommons describes a particular form of time-blinkered decision-making. If the commons effect describes the neglect of a shared resource, the uncommons effect describes a neglected *shared risk*.

It's not the only time people have risked a tragedy of the uncommons either. A couple of decades beforehand, a group of US scientists and military officials stood at another turning point.

Before the first atomic weapons test in 1945, scientists at the Manhattan Project performed calculations that pointed to a chilling possibility. In one scenario they plotted out, the heat from the fission explosion would be so great that it could trigger runaway fusion. In other words, the test might accidentally set the atmosphere on fire and burn away the oceans, destroying most of the life on Earth. They were fairly sure it wouldn't happen, but could not be 100 per cent certain.

The day of the Trinity test finally came, and officials decided to go ahead. When the flash was longer and brighter than expected, some of the team thought the worst. One of those was the president of Harvard University whose initial awe rapidly turned to fear. 'Not only did [he] have no confidence the bomb would work, but when it did he believed they had botched it with disastrous consequences, and that he was witnessing, as he put it, "the end of the world",' his granddaughter Jennet Conant told the *Washington Post*.[12]

While Wiener coined the 'tragedy of the uncommons' primarily to describe attitudes to extremely rare potentially catastrophic events like the NASA or Manhattan Project episodes, the psychology underlying the effect can describe various instances where people make poor judgements about unfamiliar future events.

Every day, lots of apparently rare things happen – and if you had asked people about their likelihood beforehand, they would have seemed very low probability. Every news bulletin is a collection of rarities: the early death of a young celebrity, a building collapse, a particularly awful crime or a person's remarkable good fortune. And every once in a while, there is a total outlier that can shape history, a so-called 'black swan'.

So, why exactly does this faulty perception of rare events happen? One of the psychological biases to blame is the 'availability heuristic', where people's judgements and decisions are based on what's readily available and most prominently in memory. As the example of Kent Cochrane's amnesia showed us in the last chapter, the mind stitches a tapestry of possible futures from experiences remembered from the past. What this means in practice, though, is that we are more likely to make predictions and decisions about upcoming scenarios based on the experiences that are available to us.

On top of this, there are a couple of other related psychological effects at play. The first is 'normalcy bias', which leads people to disbelieve or ignore warnings of future dangers because their experience suggests disasters only happen to other people, not them. And the second is 'salience bias', which describes how the 'loud, local and urgent' distractions become the predominant evidence we use to make decisions about the future. Some events, memories and experiences push themselves to prominence, overriding quieter and more subtle evidence.

Together, these biases play out commonly in daily life. The mind likes to build patterns to help predict the world. For example, if you have a number of close friends who experienced side effects of a vaccine, if your bus was late three days in a row, or if you've met three people called Jeff in your life and they were all obnoxious, then these experiences will skew your prediction. You'll expect to get side effects, that tomorrow's bus will be late and that the next Jeff will be a jerk.[13]

These mistaken expectation effects also explain why people misjudge the likelihood of rare events in the future like winning the lottery. The odds of winning the UK jackpot is 1 in 45 million, and for the EuroMillions it's around 1 in 140 million. So it's exceedingly unlikely for you to win it, but on a regular basis, *someone* wins, and their good fortune is often highly visible – champagne corks, tears and expensive purchases. The psychological availability of that person's win skews people's perceptions of their own chances. When they fill in their ticket, the memory is front of mind.

Similarly, most people have salient memories and experiences of dangers they have seen regularly on the news: terrorism, violent crime, plane crashes and so on. With emotional weight and powerful imagery, these are more likely to come to mind to create a pattern. In the weeks after a terrorist attack, for example, fears may rise of facing the same fate, even though other threats – a heart attack or car crash, for instance – are far more likely. This misperception is sometimes called 'mean world syndrome', to describe the disproportionately negative perceptions people acquire by immersing themselves in bad news on television and social media.[14]

By contrast, people are far less likely to anticipate and prepare for a rare event when there's no memory to draw on at all. The NASA and Manhattan Project officials did not pay much heed to the potential for disaster because of the low odds – but there was more to it than that. The trip to the Moon and atomic weapon test were both firsts, so there was no memory, no newspaper headline, not even a description in a history book. As the US economist Thomas Schelling once quipped: 'One thing a person cannot do, no matter how rigorous his analysis or heroic his imagination, is to draw up a list of things that would never occur to him.'[15]

The upshot? To think accurately about events in the longer term, we have to consciously look beyond the most vivid or salient memories from the recent past. Expanding the mind to the timescale of decades or centuries requires us to make a mental shift: acknowledging that what's in front of our noses can have a disproportionate effect on our choices.

Again, we'll return to 'how' later in the book, but for now let's continue our tour of psychological habits, and how they influence our perceptions of time. We need to investigate two more factors – *scale* and *pace* – which we can begin by exploring the link between the *Exxon Valdez* disaster and Mother Teresa.

The problem of scale

In March 1989, the tanker *Exxon Valdez* went aground in Prince William Sound in Alaska, causing a massive oil spill. Afterward, a team of economic researchers were charged with calculating how to weigh up the costs. If you're going to fine a company or work out potential compensation or fines, how do you decide on a dollar sum to compensate damage to an ecosystem? It's straightforward enough to value nature if you're totting up market prices, such as the value of seafood lost to the fishing industry, but what about all the other ocean creatures and habitats that were harmed?

One way that the team approached their task was to conduct a survey of people to ask them about their concern for seabirds – and they were surprised by what they found. The researchers presented participants with three numbers: 2,000, 20,000 and 200,000 seabirds. They then asked them how much they'd pay personally to protect those birds from death. They didn't mention oil spills, but the numbers did happen to tally with the toll from known disasters. *Exxon Valdez* had killed the most.

You might expect that the average value, in dollars, would grow in proportion to the numbers of seabird deaths – that people would value 200,000 birds more than 2,000. Yet the researchers were astonished to find that people gave pretty much the same answer for the three very different numbers of birds.[16] Despite the death toll getting 10–100 times bigger, people seemed to care about the same – at least as measured in dollars.

This effect is called 'scope insensitivity', and it has been observed in many more realms of life than people's attitudes to birds. Time and again, psychologists have observed that people's concern does not grow linearly as the numbers involved get bigger, whether it is deaths in a pandemic or people injured in an earthquake. Or to put it more bluntly, empathy does not scale.

The computer scientist Nate Soares once pointed out that it's very difficult to extend feelings of empathy or compassion beyond the size of the social networks closest to us, even if the will is there. 'Billions of people live in squalor, with hundreds of millions of them deprived of basic needs or dying from disease. And though most of them are out of my sight, I still care about them . . . The problem is, my internal care-o-meter was calibrated to deal with about 150 people, and it simply can't express the amount of caring that I have for billions of sufferers. The internal care-o-meter just doesn't go up that high.'[17]

It's even harder to scale that compassion across time. If you were to ask people how much they care about the eleven billion people on Earth who will be born before the year 2100, it's not eleven billion times more concern than if you told them about one person today.

At such large numbers, there's some evidence that people's concern even *drops*. A kind of moral helplessness kicks in. If people believe they can't help others, they are demotivated to try. It's sometimes described as 'psychic numbing' – an indifference that sets in once numbers of suffering get large.[18]

It's much easier for people to feel empathy for what psychologists call 'identifiable victims' instead. This explains why charities tend to use individual case studies rather than statistics to encourage donations. It's more likely to encourage donations if you show the suffering of a single hungry child or family during a famine, rather than quoting a number saying that 10,000 people need help.[19]

You can see this effect described in two quotes, from two very different people. The first, from Joseph Stalin, who apparently remarked: 'One death is a tragedy; a million deaths is a statistic.' And the second, from Mother Teresa, who once said: 'If I look at the mass I will never act. If I look at the one, I will.'[20]

Scope insensitivity may also have played a role in NASA's decision to open the door after the Moon landings: officials likely did not picture the full scale and impact of the potential disaster that could happen. Nor did the Manhattan Project officials fully imagine the scale of human disaster that burning off Earth's atmosphere would cause. But the bias crops up more often than rare moments like these ones.

Many of the grand challenges of the twenty-first century are huge problems that affect enormous amounts of people, many years into the future, from climate change to environmental pollution to global poverty. These are examples of what philosopher-ecologist Timothy Morton calls a 'hyperobject', entities of such vast temporal and spatial dimensions that they cannot be imagined with our individual mental faculties.[21] They 'out-last and out-scale us in the here and now,' Morton writes. Try to picture them in the mind's eye, and you can't.

Such incomprehensibility leads to paralysis, apathy or worse, ignoring future consequences altogether. Scope insensitivity therefore underlies some of the biggest, most complex issues of our time.

Compounding this view is a psychological effect described as the 'blamelessness of unintentional action'.[22] People often tend to judge clear, intentional transgressions as worse than unintended ones, which is why most people tend to feel it's wrong to push a person off a bridge to save five in the famous trolley problem. But when the chain of consequences for an act is obscured across long periods of time, then it's difficult to feel intuitively culpable.

However, if *scale* fosters time-blinkered behaviours, then so does the *pace of change*. We encountered one manifestation of this problem earlier in the book, when looking at the 'slow burns' of politics, but it goes deeper than that. We can see why by examining a curious black substance found in a cupboard in Australia, which would become one of the world's slowest – and most infuriating – science experiments.

The Pitch Drop Effect

One day in the early 1960s, the physicist John Mainstone was poking around a laboratory at the university where he worked, and came across something extraordinary. Stored away was a glass funnel, supported by a tripod, containing a black material. It looked like the upper half of an hour glass, but rather than sand, it contained tar pitch, an oily material that was once heated up and spread on the hulls of ships to seal them. At room temperature, the pitch seemed frozen in place.

Asking around, Mainstone discovered that the contraption was older than him by around eight years, created in 1927 by the physicist Thomas Parnell, as a demonstration for his students. Parnell had wanted to show that, while pitch may appear solid – and can be broken with a hammer at room temperature – it actually flows like a liquid. Just very, very slowly: it is 100 billion times more viscous than water. Every so often, a small drop falls from the stem beneath the funnel, but it has only happened nine times.

Little did Mainstone know that the Pitch Drop Experiment would inspire, fascinate and confound him across his entire life – a story he would recount to the Australian journalist Trent Dalton when he was an old man.[23]

John Mainstone and the Pitch Drop Experiment in 1990.

Mainstone had missed the first five drops, so by the time 1979 came around, he was desperate to see it break off and plummet. He had seen the signs that a fall was pending: the pitch develops visible fibres like tiny stalactites. But as the working week ended, nothing had happened. He popped in on Saturday, willing it to drop, but eventually decided to go home to help his wife with chores around the house. On Monday morning, he was dismayed to find a new drop below the funnel.

In 1988, he saw the signs again, so stayed close by to watch it as much as he could. One day, he got thirsty, so popped out for five minutes for a drink. In those 300 seconds outside the room, the drop fell.

So, when the eighth drop was imminent, in 2000, he and colleagues decided to set up a twenty-four-hour camera. The drop fell, and Mainstone, learning of the news via email, was delighted.

But a few hours later, another note arrived in his inbox. It began with the words 'Oh no . . .'

Mainstone died in 2013 without ever seeing the crucial moment, eight months before the next drop fell – this time, finally caught on camera.

On the surface, it may seem like a story of disappointment. But for Mainstone, the drops were not the only reason the Pitch Drop Experiment mattered to him. It represented something far more than just the moment of action when the drops fall, but rather an object that speaks to the human relationship with time and slow change: imperceptible, creeping, invisible to the eye.

'We are looking at something which takes us far beyond the way in which time is just ticking away,' he told Dalton a few months before he died. 'It moves at its own pace. We have this idea that we ought to be able to control everything. But this is not a controlled experiment. We try to make things that obey our design rules and everything else. This is not in that category at all. This is something very, very different.'

It's human nature to be drawn to the action – the discontinuity when everything changes – but the experiment illustrates that what seems permanent to human eyes is rarely so. This is a theme that has captured people's imagination for decades, from scientists to creatives. For example, the artist Julie Mecoli made a whole series of small sculptures inspired by the Pitch Drop. In her 'Dark Matter' series, she crafted a similar kind of bitumen into the shapes of cities, such as the skyline of London, Barcelona or New York. They start out intricate and detailed, but over time, imperceptibly form a shapeless mass that creeps into beakers or over surfaces.

The Pitch Drop Experiment also speaks to something important about our psychology: the difficulty we have with registering very slow changes in the world, whether it is in cities, the natural world or our own lives. How so?

It's often been observed that human beings struggle to notice slowly moving transformations, both within ourselves and in the wider world.

The best-known metaphor is *boiling frog syndrome*, the assertion that a frog won't notice the water around it heating up. However, this is an imperfect (and much-maligned) trope, because it's not actually what frogs do – scientists have tested it with real ones and they jump out. It also doesn't capture the way that slowly moving things can occasionally accelerate unpredictably, like the sudden slip of two tectonic plates, the exponential spread of a virus, or the societal overhaul of a revolution.

I'd suggest we therefore need a new term to help us describe slow change: the *Pitch Drop Effect*. This is the perception that the status quo will extend indefinitely into tomorrow, when actually it is evolving imperceptibly and could transform at any moment. The reality is that many things in the world seem like they are constant, but they are simply moving too slowly to register – until they suddenly shift, like the plummeting drop.

Take global warming. Because it happens on the timescale of years and decades, it's hard to register that the world is heating up until something serious happens like weird weather, a flood or wildfire. Referring to people's cognitive failure to note these long-term changes in the environment, the psychologist Daniel Gilbert once quipped that if invading aliens wanted to exterminate our race, they wouldn't send ships, they'd invent climate change.[24]

It also applies to social change: a long-term societal shift like growing inequality or gradual rises in local pollution often go unnoticed by the privileged.[25] The sociologist Rob Nixon once coined a term to describe these harms: 'slow violence'. Unlike more vivid forms of violence, this is a version that plays out over years or decades. It occurs 'gradually and out of sight, a violence of delayed destruction that is dispersed across time and space, an attritional violence that is typically not viewed as violence at all', writes Nixon.[26]

That is, until the communities subject to slow violence – who can see its impacts very well – have had enough. Then it can spill out unpredictably into social unrest, protests or events like riots, leaving those in power scratching their heads about why it happened. Much of the political turmoil of the past decade has been fuelled, at least in part, by the creeping disintegration of people's well-being, income, employment prospects and social structures.

The Pitch Drop Effect therefore helps to explain why the politicians we encountered in Chapter 3 tend to ignore slow-burn problems: they fail to note that harm is happening and trouble lies ahead, until it becomes a crisis. Again, here the biases of availability and salience play a major role. It's hard-baked in human evolution that we will respond more to vivid 'hot' events rather than 'cold' abstract changes – after all, this instinct helped our ancestors survive when in danger.

However, there is a particular pernicious form of slow change that we haven't yet described, and which is even harder to spot. It's different to the Pitch Drop Effect, with its periods of slow creep and sudden lurches. In fact, it's so hidden that many people live a lifetime without knowing it's there.

In the mid-1990s, the fisheries scientists Daniel Pauly and Villy Christensen were discussing the Kattegat, a shallow sea off the coast of Denmark. It connects the Baltic and North seas, and in the 1970s was one of the first places to be designated a marine 'dead zone'. Fish stocks there have plummeted.[27] Christensen, however, could recall that his grandfather once fished there regularly for mackerel, and notably had expressed irritation at the way giant bluefin tuna would get tangled in his fishing nets.

For the scientists, it was difficult to imagine a time of such abundance: mackerel could still be found, but tuna had been totally absent in the Kattegat for around half a century. It struck Pauly that many other anecdotes like this existed – old stories and memories that were not captured by the scientific data that he and his colleagues were working with. He recalled that in the 1984 book *Sea of Slaughter*, the environmentalist Farley Mowat had described the experience of moving to live on the Atlantic seaboard of Canada – a part of the ocean he had known since childhood – and coming to a disturbing realisation.[28]

'Now the sea was sounding a sombre and warning note,' Mowat wrote. He was left with 'the uneasy impression that the once familiar richness and diversity of animate life I had known in the oceanic world and on its landward verges were diminishing. There was a perceptible reduction in the numbers of seals, seabirds, lobsters, whales, porpoises, foxes, otters, salmon, and many other such whose presence I had come to take for granted.' He tried to persuade himself it was transient, but his own notes going back decades and conversations with locals could only offer a 'grim confirmation for my intimation of unease'.

Reading accounts like these, Pauly realised that he and his colleagues in ocean science were not accounting for these changes: sure, they could potentially find it in the long-term data if they looked at it, but there was something more subtle going on. As each generation entered the profession, they seemed to be accepting as normal the ocean as they saw it. Pauly called it *shifting baseline syndrome*.[29] 'Every generation begins its conscious life by assessing the state of the world and society around it and using what it sees as a baseline,' he wrote. 'However, the baselines of previous generations are commonly ignored, and thus the standard by which we assess change also changes.'

The idea of shifting baselines was a concept that would catch on and spread way beyond fisheries. A few years after Pauly proposed his new term, the psychologist Peter H. Kahn Jr of the University of Washington observed a similar effect in a completely different setting: the Black communities of Houston, Texas.[30] Kahn was curious about what children thought about the environment they lived within.

At the time, Houston was one of the most air-polluted cities in the US. In interviews, Kahn found that the children were perfectly capable of describing what air pollution was, and that other cities suffered from it, but they were oblivious to the fact that their own community was so polluted. Since that's all they had ever known, it was invisible to them. Kahn called it by a different name – *environmental generational amnesia* – but the underlying psychology was the same as shifting baseline syndrome. A generational change happening too slowly to notice.

Since then, researchers have found myriad examples of shifting baselines, in communities all over the world, from Yorkshire to Japan to the

Amazon. Whether it's changes in garden birds, large mammals, or the composition of forests, slow changes have gone unnoticed time and again. By studying people's social media posts over time, researchers have found that people have gradually normalised the increasing frequency of weird weather events and rising temperatures too.[31]

When there's a shifting baseline involved, it can be particularly hard to spot slow change. It is not just that people fail to notice it as it plays out. The problem is that a new generation comes along and accepts what it sees in the world, so never even questions it. The problem with shifting baselines is that humanity itself comes to collectively forget the world as it once was, and so makes judgements about its future based only on what it perceives in the present.

'The arc is a long one'

Slow change isn't all harms and downsides, however. For starters, identifying it in the world can serve as a focal point to connect generations. Kahn and his colleague Thea Weiss, for example, highlight the benefits when young and old people talk about the slower changes they have observed in nature. If this is done experientially, even better. It doesn't need to be the romanticised ideal of visiting a rare wild forest or hiking into inaccessible wilderness – but 'one small interaction with nature at a time'. So, it could be as simple as a grandparent and a child walking along the edge of a body of water, identifying berries on a summer's day, or even simply lying on the grass or earth – sharing memory and experience. They call this practice an 'interaction pattern'.[32]

And there are other upsides to noticing slow, long-term changes – for one thing, it reveals that there have been myriad improvements too, which our shifting baselines tend to hide from view. For example, the majority of people today accept that women have the right to vote, that racism is abhorrent and that cruelty to animals is wrong. These shifts in public opinion were hard-won, and it should not be forgotten that they were once views held only by a small, dedicated few who were willing to embark on a gradual, multi-decade effort to influence hearts and minds.

Those seeking to make the world a better place might therefore take encouragement by what progress has been achieved over the long term. Each generation faces hardships and injustices, the scale of which can feel daunting to overcome. Solving them may take a lifetime, but history suggests gradual improvements *are* possible, even if they are rarely felt in the present. Focus only on the 'doomscroll' story of the world that plays out on social media or TV, and you might well despair. There is certainly cause to worry about the world's darker trajectories, but we might derive the energy to tackle those problems by celebrating the slow victories.

Consider the words of the US Christian minister Theodore Parker, who lived in the 1800s and called for the abolition of slavery: 'I do not pretend to understand the moral universe,' he wrote. 'The arc is a long one, my eye reaches but little ways; I cannot calculate the curve and complete the figure by the experience of sight; I can divine it by conscience. And from what I see I am sure it bends towards justice.'[33]

This long-term attitude to progress is itself a baton that has been passed forward between generations. Parker's words would be later paraphrased by Martin Luther King Jr: 'the arc of the moral universe is long, but it bends toward justice'. And later, again, by the US's first Black president, Barack Obama, who referenced these words in his first election victory speech, citing what can happen when Americans 'put their hands on the arc of history and bend it once more toward the hope of a better day'.

To be truly long-minded, then, is to learn to seek out and identify slow change – and to have faith that it can be harnessed, too.

Time perspectives

By now, you may be wondering: are some people better equipped to overcome their temporal biases than others? The answer is, tentatively, yes. Everyone is subject to habits and invisible influences in their thinking, but the research does at least suggest that people have different 'time perspectives', and that some are more long-minded than others.

You can get hints of those differences in studies that aim to explore people's attitudes towards the future. For example, Bruce Tonn at the University of Pennsylvania and colleagues once asked 572 people from twenty-four countries what they think about if they hear the word 'future'. The average person described a time fifteen years ahead, and beyond that the future went 'dark' for them.[34]

Tonn's work has often been quoted as a demonstration that people have a limited view of the future. But notably, a few of the individuals in that dataset *did* have a longer perspective: the answers ranged up to 200 years. One even answered 'one million years', but Tonn decided to exclude their response because it was such an outlier that it would have massively skewed the average. (Personally, I would have been interested to know more about that individual and how they think.[35])

So let's look at what the science can tell us about individual differences in perceptions of the past, present and future. To get a handle on this research and where you might sit on this spectrum, take a moment to answer the following questions:

Read each item below and, as honestly as you can, answer the question: 'How characteristic or true is this of me?' Very Untrue, Untrue, Neutral, True, or Very True?

1. I think about the bad things that have happened to me in the past.
2. Painful past experiences keep being replayed in my mind.
3. It's hard for me to forget unpleasant images of my youth.
4. Familiar childhood sights, sounds, smells often bring back a flood of wonderful memories.
5. Happy memories of good times spring readily to mind.
6. I enjoy stories about how things used to be in the 'good old times'.
7. Life today is too complicated; I would prefer the simpler life of the past.
8. Since whatever will be will be, it doesn't really matter what I do.

9. Often luck pays off better than hard work.

10. I make decisions on the spur of the moment.

11. Taking risks keeps my life from becoming boring.

12. It is important to put excitement in my life.

13. When I want to achieve something, I set goals and consider specific means for reaching those goals.

14. Meeting tomorrow's deadlines and doing other necessary work comes before tonight's play.

15. I complete projects on time by making steady progress.

This is a short version of a test called the Zimbardo Time Perspective Inventory.[36] Developed by tech researcher John Boyd and the psychologist Philip Zimbardo (the latter who you may know for staging the Stanford Prison experiment), the full fifty-six-question version of the test measures people's perspective of the past, present and future – and how this intersects with their behaviours, personalities and attitudes.[37]

It's one of various tests that psychologists have devised to understand people's time perspectives. A second is called the 'Consideration of Future Consequences' scale, while another is the 'Future Time Perspective Scale', which measures how fast people see time moving, how big and connected the future seems, and how willing they are to sacrifice in the present.[38]

What can these kinds of tests tell us? While it should be noted that they are not specifically exploring attitudes to very long time, they do nonetheless provide hints about what kinds of 'orientation' people hold.

The first six questions in the short Zimbardo test you just read concern how **past-oriented** you are. Questions 7–12 concern your **present-orientation**, and 13–15 measure your focus on the **future**. (For a longer and more accurate test that will give you scores, visit thetimeparadox.com.)

With more answers, it's possible to score people across the following five categories:

Past-Negative: Zimbardo and Boyd describe people dominated by this view as 'Smithsonians of trauma, failure, and frustration'. They have 'a generally negative, aversive view of the past . . . marked by trauma, pain, and regret', and are 'endlessly recycling the non-modifiable past despite current good times'.

Past-Positive: 'A warm, sentimental attitude towards the past . . . a glowing, nostalgic, positive construction.' People with a strong score on this measure have 'a sense of personal continuity or a stable sense of self over time, a sense of rootedness'. This is a view of life often shaped by tradition, religion and family values.

Overall, strongly past-oriented people 'tend to be conservative, concerned over maintaining the status quo . . . they do not take risks and are not impressed by new, more efficient ways of doing familiar things'. It's also a more cooperative than competitive mindset, and hence according to the psychologists is a little rarer to find in the individualist United States, compared with more collectivist societies such as China, South Korea, Guatemala or Mexico.

Next, are the present-orientations, which are more defined by 'what is', rather than 'what was' or 'what might be'. These are people who take the bird in the hand, rather than gamble on the two in the bush.

Present-Hedonistic: In short, the belief in the good life. It's all about upsides and excitement in the moment, taking risks and seeking adventure. People high on this measure are pleasure-seeking, intense and playful, and in the workplace can be the creative ones. The downside is that they are prone to avoiding hard work or forward planning, hence they do less well in study, and their 'devil may care' attitude makes them more likely to fall prey to addictions. Delaying gratification is difficult when there are salient pleasures to be won and immediate pains to be avoided.

Present-Fatalistic: The Eeyores of the world, defined by 'a fatalistic, helpless, and hopeless attitude towards the future and life' and 'a belief that the future is predestined and uninfluenced by individual actions, whereas the present must be borne with resignation because humans are at the whimsical mercy of "fate"'. This is a view of oneself as a pawn in someone else's game. They also do less well academically, and are more likely to encounter mental health problems. On the positive side, it comes with a belief that luck can change circumstances. If they have any view of the future, it is a negative one, living in expectation of negative outcomes.

Finally, the **Future** orientation. This is a positive view of the path ahead, 'characterised by planning for and achievement of goals' and 'if–then reasoning, probabilistic thinking, logical analysis'. People high on this measure 'are willing to save for a rainy day, accept that a stitch in time may save unnecessary work later', and as such are better at delaying gratification. They tend to be health-conscious – they floss and eat well – and do better academically. There's also evidence that they are more likely to adopt pro-environmental behaviours and attitudes.[39]

There are downsides to being highly Future-oriented though, especially if a person also scores low on the present-oriented measures. It means they are less able to live in the moment, enjoying life's pleasures, and therefore can become prone to workaholism, anxiety and mid-life crisis.

Generally, people do not come out as singly categorised, rather their scores on each measure build an overall profile. The ideal mix, according to Zimbardo and Boyd, is to score low on the negative measures that encourage rumination or paralysis, and moderate-high on the others. Or more specifically, you want to avoid being too Past-Negative and Present-Fatalistic, and embrace a mix of Past-Positive, Present-Hedonistic and Future. This allows an individual to learn from their roots, live in the moment and plan for tomorrow.

Since the test was first published, there have been modifications – particularly within the Future category. When other people were tested around the world, new perspectives emerged. For example, in Sweden, psychologists identified an additional **Future-Negative** category, marked by the anxious and pessimistic belief that the future will be grim.[40] Doomist, in other words. It would be found elsewhere too. Meanwhile, Zimbardo and Boyd also added another future category: **Future-Transcendental**, measured by a different scale of questions. This perspective is associated with the religious expectation of an infinite, timeless future in heaven – the kind of eternal long view we explored in Chapter 1.

The good news is that people are not necessarily fixed in their scoring across the categories, and can change. One group of psychologists gave people mindfulness training, for instance, and found that it fostered a more balanced time perspective with less rumination.[41] And others have found that coaching interventions similar to the 'perspective taking' we discussed earlier can boost the Future perspective: for example, asking people to write down what they would want said about themselves in their funeral eulogy, or to imagine being ninety years old, sitting in a rocking chair and remembering their life.[42]

But as we'll see next, such interventions are not the only way that people's perceptions of past, present and future can be positively influenced by external pressures. To understand why, let's move on from time perspectives, and return to biases.

In 2017, the psychologist Bettina Lamm and colleagues invited children from two very different backgrounds to participate in an experiment.[43] The first group were fairly typical Western middle-class children, who lived in Germany. The second were members of the Nso people, one of 200 ethnic groups in Cameroon. This society of rural farmers is characterised by a clear social hierarchy that children are expected to integrate into from an early age. Elders are respected and obeyed, there is solidarity among peers and the mindset is more collectivist, rooted in interdependence and communal obligations.

Lamm and colleagues presented each set of these children with a treat, a marshmallow, but told them they'd get more if they waited. This was the famous 'marshmallow test', a scientific experiment first conducted in the 1960s, where researchers pose kids with a dilemma: ignore a short-term pleasure – eating the tasty marshmallow in front of them – and they'll get more later. You may have seen the test repeated on a YouTube video or TV series, where children are filmed by hidden cameras, sniffing and poking the marshmallow while they try to resist its lure. Many of them fail. (Adults may think they'd be immune to such effects, but if that were true, they would need to demonstrate that they had never eaten unhealthy foods, made rash purchases, or taken unnecessary risks.)

The marshmallow test has well-established shortcomings, particularly over claims that it holds predictive power about a child's prospects in later life.[44] But what Lamm and colleagues did with the experiment was nonetheless noteworthy, because of what it suggested about how culture could influence present bias.

Why? They found that the Nso children were far more capable of delaying gratification than the German ones.[45] Almost 70 per cent of the Nso four-year-olds waited, compared with only 28 per cent of the German kids.

The result raises a tantalising question: could societies that emphasise collectivist principles be more long-minded than individualist ones? While findings such as these should be treated with caution, they nevertheless dovetail with other research about the differences between societies – and not just among children.

For the past few decades, the psychologist Geert Hofstede sought to understand how national culture shapes people's behaviour, choices and attitudes – which he calls 'the collective programming of the mind'. Across multiple surveys, he measured six dimensions, one of which is a society's *long-term orientation*, which can be defined on a practical level as 'the fostering of virtues oriented towards future rewards, in particular, perseverance and thrift', but also more broadly as how a culture orients itself in time. According to the Hofstede approach, 'every society has to maintain some links with its own past while dealing with the challenges of the present and future, and societies prioritise these two existential goals differently'.[46] So, it's not dissimilar to Zimbardo and Boyd's Time Perspectives, but at a national scale.

Sure enough, individualist countries like the US or Australia score far lower on Hofstede's measure of long-term orientation than collectivist societies like Japan, China or Russia.[47] So, whereas the people of Japan, for example, broadly tend to 'see their life as a very short moment in a long history of mankind', Americans are shaped by ideals of 'liberty and justice for all' that, while positive in many ways, may also have allowed loosely regulated, short-termist business norms to develop in the name of market freedom. And while the US may seem a future-focused society through its technology, Hofstede argues that it is more conservative and past-facing in its social attitudes, 'preferring to maintain time-honoured traditions and norms while viewing societal change with suspicion'. These descriptions are very broad brushstrokes of course – maybe even bordering on stereotype – but Hofstede nonetheless argues his scoring can be reflective of a country's dominant cultural norms and habits.

Fascinatingly, there's even one rather speculative theory that a culture's long-term orientation could be related to how its ancestors grew crops in the pre-industrial age.[48] Before the sixteenth century, regions with agricultural and climatic conditions that led to higher yields have tended to go on to be long-term oriented in their culture and language. Plentiful surplus, goes the theory, enabled forward-thinking.

Now, depending on your point of view, these insights can be taken positively or negatively. If you believe that culture is immutable, then it would suggest that so too is a society's short- or long-term orientation.

However, there is cause for optimism. For one, it suggests that short-termism is not innate to human beings, but is culturally influenced. And over centuries, cultural values at the national scale can and do change.[49] But another reason? A growing body of evidence shows that psychological nudges applied to individuals can overcome the influence of their surrounding cultural milieu, helping them to think longer term.

A second marshmallow experiment demonstrates how. On the surface, the results of the German vs Cameroonian marshmallow study look like bad news for Western children. But in 2018 another group of psychologists, studying children in the US, found that the bias towards short-term gratification could be reduced by applying smart social nudges. In their study, some of the boys and girls were told that they belonged to a peer-group – all wearing the same T-shirt. Before the treat was presented, the kids were also informed that members of their group had waited for the marshmallow. In an apparent attempt to fit in, these children were subsequently more likely to exercise self-control than children who had not been placed in the in-group.[50]

For adults, too, the urge to align with peers can be a powerful force. Such social effects are evident in a deep well of research that shows that people often modify their behaviour to conform with others – even if they don't realise it.[51] For example, if people are told other members of their community are using less energy than them, they will tend to reduce their own use to align with the norm.[52] The same nudges have been successfully applied to encourage other pro-social acts that ask us to think beyond the present, such as healthy eating, paying tax, or even organ donation.[53] This would suggest that if your peer group demonstrates longer-term attitudes, then you are more likely to embrace those values too. In short, the long view has the potential to spread.

But of all the social nudges that promise to foster long-mindedness, perhaps the most intriguing is a concept called 'intergenerational reciprocity', which encourages people to think about their social relationships across time too. This body of research suggests that when people are asked to think more deeply about the norms and traits of their ancestors and descendants, then it influences their own behaviour to act more benevolently long term.

This work began with an experiment where people were asked to imagine themselves as the retiring CEO of a commercial fishing business. The CEO, they were told, needed to decide how much stock to leave to their successors. Notably, if people were told that past generations had been generous, they were more likely to do the same.

In a different study, people were bequeathed money, and asked to decide what to do with it. Again, people were more generous to their own descendants if they were told their forebears had been, even if it cost them. These were just laboratory experiments, but they hint that emphasising a sense of communal stewardship is a powerful thing indeed.[54]

Later, psychologists were interested to see if this sense of stewardship could be more generalised, so conducted a study where they asked people to reflect on the following question (simplified for brevity):

'In what ways have the sacrifices made by your parents, grandparents, or great-grandparents allowed you to live the life you lead today?'

People who did so were more likely to express a feeling of moral obligation to their descendants and future generations in regards to climate change, compared with another group who were simply asked to reflect on their grandparents' fashion choices.[55]

And in another study, psychologists asked people how they want to be remembered by future generations – so-called 'good ancestors'. They were subsequently more likely than others to express pro-environmental attitudes.[56] 'Prompts that encourage people to think about how they would want to be remembered (or perhaps what they don't want to be remembered for) may effectively promote environmental behaviour by framing decisions as "win–win" for both present and future generations,' the researchers concluded.

Finally, psychologists have observed one more intergenerational reciprocity effect by asking people to imagine the traits of future people too. In one study, researchers asked 600 people to imagine the year 2050, and picture a world where various policy changes had been made, related to climate change, for instance, or drug laws.[57] They found that, if people were asked to imagine those future generations as having the attribute of 'benevolence' – a warmth and sense of moral duty – those people were more likely to support a policy or personal behaviour change that would lead to this future in the present.

The researchers suggested that, if politicians or activists wanted to encourage people to act on an issue like climate change, they might tailor at least some of their messaging to focus on reciprocity and benevolence, rather than only warning of harms and dangers. 'Overall, people appear more motivated to act in support of creating a society with better people (warmer and more moral), than they are motivated to act in support of creating more favourable societal conditions for people to live in,' the scientists concluded.

So what do these studies tell us? Collectively, they suggest that when people are asked to mentally time-travel to reflect on the traits of their ancestors or the attributes of their descendants, it can create a social nudge that encourages a longer view.

Ultimately, humans are social beings, and it is through our relationships and communities that we transcend our baser instincts and habits. The mind may be prone to cognitive biases, but across history, people have created and embedded positive social norms that have proved to be stronger. While culture has the power to shorten perspectives, it's also possible for it to extend them too. In Part III, we'll fully explore this idea from various angles, by profiling some of the world's most long-minded people and communities, who have forged stronger links with their ancestors and descendants, and found ways to collectively extend their view into deeper time.

But before we do, there's one more quirk of the mind's perception of time that deserves scrutiny – it concerns how *language* can influence people's attitudes to the past, present and future. It turns out that the specific words we speak have the potential to either shorten or lengthen our view too.

To discover why, let's begin on the side of a valley in Papua New Guinea, where we'll meet a group of people with a perspective of time unlike anyone else's on Earth.

6

Long-terminology:
The Power of Language

'The soul when thinking seems to me to be just language.'

PLATO

One day, on the side of a valley in Papua New Guinea, the cognitive scientist Rafael E. Núñez and his colleagues were in a conversation with a man called Danda, who had a very unusual way of thinking about time. The more that the man spoke, the more they realised that Danda didn't perceive the past and future like them. For him, time flowed uphill.[1]

Danda is a member of the Yupno indigenous people, who live in the mountainous interior of the island of New Guinea. The landscape features forested hillsides, steep grassy slopes and ridges. There are no tarmac roads, nor electricity.

Núñez was interviewing people in the Yupno village of Gua to ask them about their perceptions of the past and future.

'What is the difference between yesterday and tomorrow?' he asked.

Danda paused and thought for a moment. When he answered, he made telling gestures. Describing yesterday, he waved his hand backward down towards the base of the valley, and as he mentioned tomorrow, he gestured upslope.

When Núñez and colleagues repeated the question facing the opposite way, Danda switched hands – still gesturing downhill to the past and uphill to the future.[2] For the Yupno, time is understood *topographically*.

Núñez and colleagues suggested that the Yupno understanding of time could be based on the group's long-term history: their ancestors arrived from the sea and began in the lowlands before climbing up to settle in the mountain valley at an altitude of 2,500 metres (8,200 feet).

But it got more complex when they conducted more interviews. Inside the Yupno buildings, the community has a slightly different topography for time. While the floor may be flat in their homes, people gesture towards the door when referring to the past and away from the door when talking about the future.[3] This applied no matter which way the house was facing on the valley sides. It was still a way of understanding time as space, but the cognitive framework had been subtly remapped indoors.

The Yupno are one example of many from around the world where people have developed slightly different conceptions of past, present and future through the language they speak. These examples raise a tantalising question: could the language of time influence how we think about it?

This is an idea that has occupied intellectuals for centuries. The philosopher Immanuel Kant wrote of the entwined relationship between language and thought, observing that 'thinking is speaking with oneself'.[4] And more recently, psychologists and linguists have picked up the baton. Some of them have even suggested that language might be seen as the 'instruction of imagination'.[5] To what extent could this be true?

As we learned from the fall of Icarus and near–far effects (see pages 121ff.), people tend to conjure up a landscape to imagine time. This feeds into the words we use: 'The last chapter is *behind us*, so let's *look forward* to what's next.' It's also common for people to use spatial metaphors. In English, time can be a physical object: you can have time on your hands, or you can give it to someone. Or it can be a container: you can fill it with activities. Or it can be volume itself: oceans or sands of time. One of the more unusual spatial metaphors in English is a 'whale of a time', equating the concept of large with an enjoyable duration.

This way of making time tangible and spatial is deep-rooted and entrenched across many cultures. But different languages have specific vocabulary and metaphors.

As we've seen, English speakers tend to talk about duration in terms of distance or length. The English title of this book is, after all, *The Long View*. Yet other languages, such as Greek, talk about duration in terms of amount or quantity. In Greek, the word *'makris'* means 'long' and is used to describe things like ropes, roads, arms and so on. But a Greek speaker would be more likely to describe a long meeting, night or relationship using the word *'megalos'*, which in spatial contexts means 'physically large'. And in Spanish, the direct translation of 'long time' – *'largo tiempo'* – sounds awkward, so *'mucho tiempo'* (a lot of time) is preferred.

Greek and Spanish do share other commonalities with English though. In most European languages, the past, present and future are visualised horizontally, with time's arrow pointing towards the right. If you give an English speaker a series of photos, showing a person at different ages in their life, and ask them to arrange the pictures, they will most likely put the youngest person on the left and the oldest person on the right.

Some languages have different orientations, which mirrors their writing direction. Given the same photoset of ageing people, Hebrew speakers would tend to lay out the cards from right to left. And Mandarin speakers, whose writing runs from top to bottom, often visualise the past as 'above' and the future as 'below'. When you ask people to use hand gestures to demonstrate 'next month' or 'yesterday', Mandarin speakers tend to use the vertical axis. Earlier events are described as *'shàng'* or 'up', and later events are said to be *'xià'* or 'down'.[6] (The word *'shànyuè'* means up-month or 'last month', while *'xiàyuè'* means down-month or 'next month'.)

Vertical language for time is present in English too, but only occasionally. One form of words that's particularly relevant to the long view is the idea of 'handing down' knowledge, stories, legacies and responsibilities from generation to generation.

The speakers of Kuuk Thaayorre, spoken by the Pormpuraaw Aboriginal community in Australia, have a particularly fascinating way of orienting past, present and future.[7] At all times, the Pormpuraawans know which direction they are facing – north, south, east or west – and they use these directions descriptively (for example, 'she waved her east hand'.) Instead of saying hello, a Kuuk Thaayore speaker asks 'where are you going?' And the response usually involves a compass direction.

When asked to arrange photos of ageing people, Kuuk Thaayorre speakers placed the photos in a line running from east to west. Their chosen arrangement differed depending on which orientation they were facing at the time.[8] 'It was a beautiful pattern,' writes Lera Boroditsky, a US-based psychologist who studied the community's way of thinking. 'Pormpuraawans think about time in ways that other groups cannot, because they lack the necessary spatial knowledge. Many Americans could not lay out time in absolute coordinates even if they wanted to.'[9]

Indeed, Boroditsky herself (who was born in Belarus and lives in the US), found herself embarrassed when the Aboriginal community members asked where she was from. She could not describe the direction, nor point the right way. Which way was California from Australia? She wasn't sure.

Like the Yupno example of Papua New Guinea, you can also find intriguing differences between how cultures see themselves three-dimensionally within time. For English speakers, the past is at their back and the future is something to look forward at, as captured by the phrase: 'The worst is behind us, and the best is still ahead.' This is the case in the majority of Indo-European languages, as well as others such as Japanese or Hebrew.

Not so for the Aymara people, an indigenous group in South America. A few years before Núñez visited the Yupno villages in Papua New Guinea, he and colleagues had travelled to the Andean highlands that span Peru, Bolivia and Chile, where they had several conversations with members of the Aymara.

The Aymara language describes the future as 'behind', and when talking about the past, speakers gesture in front of them.[10] So, the Aymara word for past is '*nayra*', which means 'eye', 'sight' or 'front', and the word for future is '*q'ipa*', which translates as 'behind' or 'back'. 'As such, Aymara speakers tend to speak more often and in more detail about the past than about the future,' writes Núñez. 'Indeed, often elderly Aymara speakers simply refused to talk about the future on the grounds that little or nothing sensible could be said about it.' (Notably, younger members, who also speak Spanish, tend not to share their elders' spatial–temporal perspective.)

Aymara is not the only language with the past–front/future–behind orientation. The same can be found in Malagasy in Madagascar, where past events are described as 'in front of the eyes'. It makes logical sense for the concept to have emerged in more than one place if you consider the fact that the past is known and seen, whereas the future is not.[11] A Malagasy speaker told one researcher that future events should obviously be behind, because 'none of us have eyes in the back of our head'.[12]

The Toba language of Bolivia has a past–future orientation that seems particularly complicated at first glance, but is remarkable in its logic. The recent past begins directly in front, but then *curves upward* the further back you go until eventually it is far above the head. There, it merges with the remote future, somewhere vertically above. The point is that both the deep past and future are similarly invisible. The *immediate* future, however, is directly behind the body – so a person can, in principle, turn around and look over their shoulder to see it because it is close.[13] If you were to draw that out as a diagram, it would look like a circle, with the observer at the base.[14]

Meanwhile in Tahitian, speakers say: '*I mua*' (going forward) and '*I muri*' (going backward), but when you talk about going towards the future you say: '*I muri mai*', which is pretty much: 'going well backward'.[15] There's clearly a shared Polynesian root going back centuries, as this echoes the Maori genealogical concept of facing ancestors as you step into the future: '*Ka mua, ka muri.*' And in the Hawaiian language, which derives from the Tahitian, there's a proverb: '*I ka wā mamua, ka wā mahope*', which means 'the future is in the past'.

Then there are the differences between the way that people around the world see themselves moving (or not) on a timeline. In English and various other languages, people tend to imagine themselves travelling towards the future – in this mental picture, the timeline is stationary, and they are the protagonist, striding towards tomorrow.

Mandarin speakers don't do this, however. While they use the same future-in-front, past-behind terms to talk about time (*qia'n* is front, and *ho'u* is back), a key difference is that they tend to see themselves as fixed on an imaginary timeline, with time *moving through them*.[16] So in some ways, time is more like a wind than a track or a path.

The Amazonian Amondawa group of Brazil have no word for 'time' at all, and do not speak of hours, weeks, months or years, lacking a calendar system or clocks. Their language also does not use space to denote an individual's position within time.[17] Instead they use other references such as the word '*kuara*' (Sun), to describe rough intervals and events. The start of the summer is '*o'an kuara*' (the Sun is born), and as they approach the end of the season it is '*kuara tuin*' (small Sun). The day '*ara*' is not measured over twenty-four hours, but simply three parts: '*ko'ema*' (morning), '*karoete*' (noon/afternoon) and '*iputunahim*' (night/intense black).

The Amondawa also do not measure their ages numerically, because their numeral system does not go above four. Instead, individuals change their name when they reach a new life stage. So for example, a newborn girl might start as Tape, become Kunhate as a young adult and Mytãg as an elder. One can only speculate how that might change your sense of self across a lifetime if it had been a practice where you grew up.

What to make of all this? Could these various tongues evoke different conceptions of the long view? There are no firm answers, but we might get some clues from what scientists have found so far about how language can influence thought.

It's not controversial to suggest that language, broadly, plays a crucial role in how we construct reality in our heads. Words and metaphors help us to organise and understand our messy surroundings and the abstract concepts we encounter. And each language, developed over thousands of years, offers a different mode of navigating through that world.

But to be clear, this doesn't mean that language *constrains* thought. In the mid-twentieth century, some researchers believed that if a tongue lacked a word or concept, it might prevent its speakers from thinking deeply about it. This idea was particularly associated with the Native American Hopi people, based in Arizona, who some researchers believed had no words and grammar for time. As the linguist Benjamin Lee Whorf famously asserted in the 1940s, the Hopi therefore have 'no general notion or intuition of time as a smooth flowing continuum in which everything in the universe proceeds at equal rate, out of a future, through the present, into a past'.[18]

However, it is not true. Apparently, Whorf never actually visited Arizona to study the Hopi – his argument was based on conversations with one speaker in New York City. The linguists who actually did, decades later, would show in detail that the Hopi have various ways of describing past, present and future.[19]

But even if the Hopi *did* lack temporal vocabulary, it would not necessarily have constrained their thoughts. After all, it's not difficult to find examples of supposedly 'untranslatable' words outside one's own tongue.[20] For example, as an English speaker, I lack a translation of the German '*Schadenfreude*' or the French '*savoir-faire*', but almost certainly felt those emotions as a child before I knew those words.

Similarly, some languages have temporal vocabulary that my own tongue lacks. Consider the word '*guwaya*', from the Aboriginal language Wiradjuri. It can mean 'in a little while; later; after some time', but more poetically, it has also been translated as 'still and yet and for all times . . . all times are inseparable; no time is ever over; and all times are unfinished'.[21] You probably don't speak Wiradjuri, but that doesn't mean that you can't comprehend what '*guwaya*' might mean, albeit with a bit of explanation.

Yet while it may not be true that language *imprisons* our thoughts, ideas and beliefs, it does ask us to *express* them in specific ways. When your language necessitates that you use specific details, it also means you must notice and remember them as you navigate the world. This is a subtle point, so to understand it better, consider the following story of life and near-death, told by the linguist Guy Deutscher in his book *Through the Language Glass*.[22]

Bambi and the boat

In 1980, an Aboriginal man in Queensland called Jack Bambi had been travelling on a boat to deliver clothing and other items when a storm struck. He and another man abandoned ship after it capsized and had to swim more than 5km (3 miles) back to land.

On the beach, they prayed and looked out to sea, where they saw a shark swimming within the same patch of ocean they had just escaped. They then walked for hours to the home of a local missionary, who, rather unsympathetically, sent them back to retrieve the boat.

Bambi would later recount the experience to the linguist John B. Haviland. Watching him tell the story more than once, Haviland noted down some intriguing observations.

Bambi spoke Guugu Yimithirr, which is a language that shares commonalities with Kuuk Thaayorre, describing the position of the self and objects with north, west, south and east. So, he mentioned compass coordinates throughout, coupled with hand gestures that reflected the directions. For instance, Bambi expressed how he dropped out of the boat to the *west*, and how the shark was swimming *northward*. He told the story in pretty much exactly the same way when interviewed two years later too. Even in a life-threatening situation, Bambi apparently took accurate note of all the orientations of space, so that he was able to express the story in his language.[23]

While Guugu Yimithirr is far from a commonly spoken tongue, plenty of the more widespread languages also contain quirks and features that shape how their speakers express ideas. This may have unanticipated consequences: influencing what we take in about the world, how we talk to each other about it and the other associations that we make. Possibly even our perceptions and subsequent reasoning. Or as Deutscher put it: 'What we are not sufficiently aware of is the force of the habits that language can create.'[24]

As an illustration, consider that some languages ask their speakers to be specific about gender when describing, say, a neighbour or friend. If you say in English: 'I had lunch with a friend from work today', you needn't mention if that person was male or female, but in plenty of languages, you have to. And the necessity of gender assignations in some languages may also bring other associations and habits of thought. In one often-cited study by Boroditsky and colleagues, German and Spanish speakers deployed different adjectives to describe a bridge, which is feminine in the former and masculine in the latter.[25] The German speakers used the words beautiful, elegant, fragile, peaceful, pretty and slender. The Spanish speakers used big, dangerous, long, strong, sturdy and towering.[26] (Intriguingly, the German for 'past' – *Vergangenheit* – and 'future' – *Zukunft* – are also both feminine, while in Spanish they are masculine – *pasado* and *futuro*.)

But when it comes to how language creates habits of thought, one of the more striking quirks comes via grammar – specifically, whether a language has a 'strong' future tense or not. Unlike English, Spanish or French, the languages of German, Japanese, Mandarin and most of the Scandinavian tongues have a 'weak' future tense.[27] Some researchers even go so far as to describe them as 'futureless', but that's somewhat misleading. It's not that they have no future, it's more how it is expressed grammatically. So, for example, if a person wants to tell you the weather forecast looks bad in colloquial German, it's common to say '*Morgen regnet es*' – 'tomorrow it rains', not 'tomorrow it *will* rain'. Mandarin is similar: '*Ming tian xia yu*' – 'tomorrow fall rain'.

What makes this strong vs weak difference so intriguing is that it may subtly influence how people *perceive and plan for* the future too. This is a hypothesis that can be traced back to 2013, when the behavioural economist M. Keith Chen conducted a big data analysis of people's attitudes to retirement, saving and health behaviours across many languages.[28] He compared people who were otherwise very similar in terms of wealth or status, but found that people whose first language had a weak future tense (like German) were 31 per cent more likely to have put money into savings in a given year, had accumulated 39 per cent more wealth by retirement, and were 29 per cent more likely to be physically active. They were also less likely to be smokers, and a bit less likely to be obese.

Why? In a sentence like 'tomorrow it rains', it's possible that the present-tense form of the verb makes that weather feel psychologically closer and more concrete. Therefore, it's theoretically possible that this is also how a German or Chinese speaker thinks about their future *self* too.

A similar effect can be seen in fiction. When telling stories, writers often turn to the present tense when describing events to make it feel more immediate to the reader. Charles Dickens knew this, deploying present-tense storytelling for much of *Bleak House*.[29]

It should be noted that Chen's claim is a hypothesis – he observed the effect in an analysis of big data, rather than by conducting local interviews or experiments, so it could be an artefact. Other unknown cultural effects could be at play too.[30] But if true, it would align with the psychological distancing near-and-far effects we explored in the last chapter.

And subsequent work has lent the hypothesis some support. For example, in corporations, speakers of weaker future languages display a greater willingness to save and invest in R&D.[31] It's also associated with pro-environmental behaviours, and support for future-oriented policies. Psychologists compared the environmental attitudes of Mandarin (weak future) and Korean (strong future), and found that Mandarin speakers perceived greater urgency in terms of the action required to tackle environmental problems. Another group compared Russian (strong) and Estonian (weak) speakers, and found the latter more likely to support a green tax, even after controlling for other possible factors.[32]

One of the more ingenious studies involved interviewing around 1,000 primary-school children who live in the Italian city of Merano. This was a smart place to test Chen's hypothesis because half the population speaks German, and the other half speaks Italian, a language with strong future. The children therefore lived pretty similar lives, next door to one another. The experiment was a bit like the marshmallow test, to discover how willing the kids were to delay gratification – but in this case the prizes were little tokens that they could exchange for prizes like sweets, stickers, marbles or balloons. Crucially, when they were presented with choices between, say, two tokens now, or four tokens later, the German children were more likely than the Italian speakers to delay immediate rewards and wait for the bigger prize.[33]

Again it's still possible that unknown cultural differences between the kids played a role, but another clever study in 2020 suggested otherwise. Here, the researchers interviewed individuals who speak two languages: one of German, Dutch or Mandarin (weak) and one of English, French, Spanish, Hindi (strong).[34] Amazingly, when these bilingual people were given instructions in, say, German or Mandarin, they were more likely to delay gratification in an economic game than when they were addressed in English or the other strong future tongues. In other words, it couldn't be down to cultural differences: *the same person* changed their behaviour and choices.

Sadly, there are not yet any scientific studies that I am aware of which compare how different languages influence people's perceptions of the very, very long future. It would be fascinating to explore how deeper time is understood and perceived in all these different tongues.

Still, it is nonetheless fun to speculate. It may well be that the language you speak, and the culture you live in, evoke a subtly different set of mental habits when thinking and conversing about long-term time.

As we've seen, it's possible that languages which encourage their speakers to think of themselves as closer to the future might be better equipped to delay gratification. Could these speakers therefore be better placed than others to talk and think about the long view? Now when I speak in English, I find myself wishing I had a richer grammar and vocabulary to express the future, which doesn't frame it as somewhere else, many miles away, and psychologically distanced from the present. For this, English is somewhat lacking.

I can't fix the grammar of my mother tongue, but I do believe some vocabulary choices may be better than others. For example, I find that the descriptions of a *long* or *deep* future evoke a more helpful picture in the mind than a *far* or *distant* one. The long/deep future extends from where we are now, whereas the far/distant future is geographically detached: an island across an ocean, unknowable and populated by people with irrelevant concerns. What's more, the long/deep future is also cumulatively bigger than, say, a single date in the year 3000. In short, there's more of it to care about, because it includes everything that happens between now and that year.

I also wonder if my own Western perspective of time is entangled with my language's tendency to frame me as moving through it, facing forward. If I frequently talk about myself as the protagonist carving my way through the hours, years and decades, it places me at the centre of a temporal universe – a rather individualistic view. I can only speculate about how people think in other languages, but if time is routinely talked about as flowing past the self like a wind, if the future is always discussed while the mind's eye looks backward to one's ancestors, could it foster a slightly less egocentric habit of thought?

Might there be forms of words that foster a closer relationship with past and future generations? As we'll discover in Part III, some cultures have concepts of time that defy easy translation, such as the non-linear perspective of Aboriginal Australians, roughly described as 'everywhen'.[35] This feels very different to more individualistic Western idioms like the imperative to 'make history', or technological calls to 'shape' or 'disrupt' the future to one's own ends.

Perhaps the metaphors we commonly reach for play a role too. There's evidence from psychology experiments that people's attitudes can be shaped by metaphor without them realising. For example, when crime is described to people as a 'virus infecting the city', they are more likely to support social reform and prevention policies. When crime is imagined as a 'wild beast preying on a city', people are more likely to support punitive incarceration.[36]

Studies like these make one wonder how the Western idea of time as *money* shapes habits of thought. You can spend it, waste it, buy it, borrow it, or run out of it. Some researchers speculate that the language of time-as-commodity only began to be used in the West following industrialisation in the late eighteenth and early nineteenth centuries. It was only when labour began to be paid by the day or year that time began to be understood as something which could be acquired or lost. Older texts, such as the Qur'an, do not contain commodity metaphors for time.[37]

Meanwhile, at some point in the last 200 years or so – at least judging by references in English-language books[38] – people also started talking about time as an *adversary*: the idea of beating the clock or racing against it, or as something to kill. Certain phrases, such as the idea that time is a bitch or an enemy, are even more recent.

With time framed as a finite resource or foe, it makes me reflect on a word like 'lifetime'. In English, you'd talk about what you want to achieve 'within my lifetime', rather than 'within my lifespan'. Awkward as it sounds to the English ear, perhaps the latter would be subtly better for thinking multigenerationally: after all, a bridge spans two points. But now we're getting really speculative.

Suffice to say, whatever your mother tongue, our global survey of language suggests that the ways people talk about past, present and future vary significantly across the world, and therefore could be subtly shaping perspectives beneath conscious awareness. The words we use matter far more than many people might realise.

But this should not be a cause for pessimism. After all, language is also one of humanity's greatest inventions, helping us categorise and make sense of a messy, complex and abstract world. It is also something that we can update, evolve and improve.

As psychologist Lera Boroditsky once wrote: 'One of the great mysteries of the mind is how we are able to think about things we can never see or touch. How do we come to represent and reason about abstract domains like time, justice or ideas? The ability to cognitively transcend the physical is one of the very hallmarks of human intelligence.'[39]

When our ancient ancestors began to speak to one another, it suddenly became possible to place ideas in another person's head, without needing to physically experience them. Those ideas could then be mentally manipulated, iterated, expanded – and then passed on. And all languages are continually evolving. Perhaps we are yet to find the best possible form of words for talking about very long-term time.

In the meantime, we can find ways to unlock the long view through other methods. And that brings us to Part III, our exploration of the myriad paths to true long-mindedness.

We've explored how to spot and transcend the cultural stresses and psychological habits that shorten temporal perspectives. Now it's time to expand our view into the really long term – over hundreds, thousands or even millions of years. In the coming chapters, we will seek wisdom and insights from the people, movements and organisations who are showing how to project the mind far beyond the present, spanning the worlds of philosophy, religion, indigenous culture, art and science.

This may be a challenging task, but as we'll discover, there is more than one route to the same destination . . .

PART III

THE LONG VIEW
EXPANDING OUR PERSPECTIVE OF TIME

7

A Delightful Horror:
Deep Time's Sublime Scale

'Infinity has a tendency to fill the mind with that sort of delightful horror, which is the most genuine effect, and truest test of the sublime.'

EDMUND BURKE[1]

In the 1860s, the scientist Karl Ernst von Baer tried to imagine what it might be like to have a very different awareness of time. He pictured two people, one called the Minute Man, who experienced a full lifetime in less than an hour; and the other called the Millennium Man, whose life stretched over millions of years. It was a thought experiment that holds useful lessons about our own temporal perceptions – and what we can and cannot experience directly.

Von Baer was something of a polymath as a scientist: he studied the Arctic permafrost, and would later become known as the father of embryology, discovering the mammalian egg. His willingness to embrace philosophical questions and a broader view of nature was evoked by an intriguing lecture he delivered about how time perceptions might vary within the biosphere – published as *Welche Auffassung der lebenden Natur ist die richtige?* (Which conception of living nature is the right one?).[2] He speculated wildly in the talk, and there were some things he didn't get right – proposing that an animal's heartbeat correlated with its tempo of perception, for example. But one question he posed still resonates today: what if humans perceived time running at a radically different rate?

So, he imagined a Minute Man, who would experience everything at a much faster pace to the rest of us – so fast, in fact, that his entire lifetime would last only around forty minutes in our time. To the Minute Man, our world would seem almost frozen: a bullet would hang near-stationary in the air, hovering alongside glistening raindrops. 'Every sound we hear would surely be inaudible for such people if their ear was still organised in the same way as ours, but they might instead hear sounds that we cannot,' von Baer remarked.

The second hypothetical person he imagined lived their life at a far slower, geological pace. The Millennium Man's life would be spread out over millions of years, with their heart beating once every few centuries. For an individual with this point of view, everything in our world would move by too fast to see: all living things would emerge, die and rot before he had a chance to register it. But transformations that are far beyond our own sensory faculties would play out in front of his eyes: the orogeny of mountains, rising and falling like ocean waves, the continents drifting across Earth's surface and the constellations of the Milky Way rearranging into new shapes across the sky. Species would come and go, and the grand story of natural selection would materialise as he watched.

Reflecting on von Baer's thought experiment today reveals that we have mastered one of these perspectives, but not the other. It's now fairly easy to step into the shoes of the Minute Man: through camera technology, we can slow down time in extraordinary fidelity. You can do it with your smartphone. However, the perspective of the Millennium Man is a little more elusive. Just as the Minute Man could not observe the flowers emerge in spring, see the leaves change colour in autumn or watch a child grow up, so too do our minds struggle to perceive what it would be like to take in million-year chronologies. We can simulate the isostatic uplift of mountains or the evolution of species through theoretical models, but actually *experiencing* it is beyond our sensory capabilities.

As such, it can be daunting to reflect upon the really long view. Nature measures the passage of time in stratigraphic eras and evolutionary adaptations. Civilisations measure in religions and empires. Societies in states and revolutions. Yet for a human being, the units of significance are still years, generations, or at most, our short lifespans.

Thinking about one's role within deeper tracts of time is among the most difficult concepts for the human mind to approach, so far removed as it is from daily experience. Deep time is a vast, seemingly bottomless pit, an ocean horizon in every direction, or the blackness of infinite space between the stars. The more you extend the mind into longer timescales, the more you realise just how fleeting a single life can be. It can feel like each of us are grains of sand within a river we'll never fully experience.

To be long-minded is to extend your perspective beyond salient daily experience – and even beyond your own lifespan. But how can this long view be embraced if none of us can truly know it ourselves?

The Victorian geologist Charles Lyell once wrote of the challenge of contemplating the 'immensity of time' with a mixture of melancholy and reverence, describing a 'painful sense of our incapacity to conceive a plan of such infinite extent'.[3] He lived in a period when the Earth sciences were beginning to unlock million-year chronologies, but his words seem to evoke a pious form of intellectual surrender in the face of enormity.[4]

To illustrate how he felt, Lyell described a circle of light expanding into the dark – it illuminates as it goes, but as its circumference grows, so does the boundary between light and dark. In other words, he was suggesting that the more we learn about our place within time and nature, the more we become aware of our true insignificance and how little we know. 'While the scheme of the universe may be infinite, both in time and space, it is presumptuous to suppose that all sources of doubt and perplexity would ever be removed,' Lyell wrote.

Much later, the science writer John McPhee, who helped to popularise the idea of deep time in the 1980s, suggested that human beings simply may

not be able to grasp the concept to its full extent. 'The human consciousness may have begun to leap and boil some sunny day in the Pleistocene,
but the race by and large has retained the essence of its animal sense of
time,' he wrote in his influential book *Basin and Range*. 'People think in five
generations – two ahead, two behind – with heavy concentration on the
one in the middle. Possibly that is tragic, and possibly there is no choice.'[5]

McPhee suggested that the units of years, the common currency of
humanity's temporal understanding, become ever-less useful and tractable
once time becomes very big. 'Numbers do not seem to work well with
regard to deep time. Any number above a couple of thousand years – fifty
thousand, fifty million – will with nearly equal effect awe the imagination
to the point of paralysis,' he wrote.

But let's not be quite so pessimistic.

Throughout history and around the world, there have been long-
minded people, groups and cultures who have found meaning, clarity and
purpose by reflecting on their role within deep timescales. In the coming
chapters, our goal is to learn from them. As we'll discover, there are many
routes to the long view: cultural practices, ethical lenses, artistic devices
and more. Some are carving out new insights from scientific or philosophical inquiry, while others are drawing on the wisdom that already lies
hidden within their faiths and traditions around the world. What unites
them is the belief that long-mindedness is within our grasp.

While the direct experience of von Baer's Millennium Man may always
be beyond our sensory faculties, that doesn't mean we cannot get a taste
of what it is to gaze over thousands, millions or even billions of years.
Nor does it mean that we should not try.

A sublime partnership

So, where to begin? In the opening pages of this book, I described my own
starting point in the quest for a longer perspective: my daughter's birth and
imagining her as an old woman welcoming in the twenty-second century.

This reflection would lead me to the words of the eighteenth-century conservative politician and philosopher Edmund Burke, who framed society as a partnership.

Here's what he wrote in more detail:

> Society is indeed a contract . . . It is to be looked on with other reverence; because it is not a partnership in things subservient only to the gross animal existence of a temporary and perishable nature. It is a partnership in all science; a partnership in all art; a partnership in every virtue, and in all perfection. As the ends of such a partnership cannot be obtained in many generations, it becomes a partnership not only between those who are living, but between those who are living, those who are dead, and those who are to be born.[6]

More recently, I've found it illuminating to imagine just how far Burke's partnership might extend from my own life. It is an exercise that anyone can do, revealing a personal proximity to otherwise distant periods.

So, if my daughter Grace has children at roughly the same age I became a parent, I can estimate that they will live to the 2130s. Assuming medical science continues to improve longevity and reproductive fertility, her grandchildren could reach well beyond the 2160s, and her great-grandchildren could expect to see in the twenty-third century. Longer time may be difficult to conceptualise in the mind, but I find it a little easier when I think in the units of generations.

Using the same mental shortcuts, the past is also only a few generational steps away. When Grace was born, I found myself reflecting on what new parenthood might have been like for my own mother and father, and how the world must have felt at the time. I was born in 1980, when Ronald Reagan won in a landslide and John Lennon was assassinated. My grandparents, meanwhile, held their first children in the late 1940s, in a world still reeling from the Second World War. I can trace this chain of parents and births backward more than 200 years, on my father's line, to a long-forgotten coachman and washerwoman living in southern England. Both born in the 1780s, their parents welcomed them into a decade when Great Britain was fighting the US, and the Bastille would soon be stormed across the Channel.

Marked on a historical timeline, all these world events feel far apart, but they draw a little closer when I imagine my forebears living alongside them. This mental journey to imagine my ancestors is also an exercise that fills me with gratitude. If any one of these great-grandparents had not met, I wouldn't be here.

Our cross-generational links do not end with our family ties, however. What I find particularly mind-boggling is the ancestry that you, me and everyone shares – it's far more entwined than first appears. If you could draw your full family tree, and I did the same, eventually we'd see a matched name on the chart. Go back far enough, and pretty much everyone's trees join up.[7]

We know this through straightforward mathematics. For starters, your own tree's branches must have some entanglement in the past. Since your great-grandparents double every generation, that would seem to suggest that you had 1,099,511,627,776 ancestors around 1,000 years ago. But that's impossible, because a trillion people far exceeds the planet's population at the time, which was only 200–300 million people.[8] Therefore, distant cousins must have met and married over the years, mixing up the neat branches.

'Pedigrees begin to fold in on themselves a few generations back, and become less arboreal, and more mesh or web-like,' writes the geneticist Adam Rutherford in his book *A Brief History of Everyone Who Ever Lived*. 'You can be, and in fact are, descended from the same individual many times over. Your great-great-great-great-great-grandmother might hold that position in your family tree twice, or many times, as her lines of descent branch out from her, but collapse onto you.'[9]

This entanglement also leads to some counter-intuitive conclusions. Mathematical and genetic modelling suggests that at some point as little as 3,000–4,000 years ago, there was an individual who was a common ancestor to *every human being alive today*. Someone who connects us all, no matter where we live, what colour of skin we have, or who we are.[10]

That this person lived so recently might seem difficult to believe. But as Rutherford points out, 'we're not very good at imagining generational time. We see families as discrete units in our lifetimes, which they are. But they're fluid and continuous over longer periods beyond our view, and our family trees sprawl in all directions.'

This enmeshing will continue into the future, too, and may intensify given the rise in global migration. Your lineage may well die out, but if it doesn't, you could stand to feature in the shared ancestry of millions of people living in the future. Perhaps one day, as your deep descendants meet and entangle their family trees, you might even become a great-grandparent to large swathes of humanity.

You don't need to have a child to play a role in this story. Every day, we each make decisions that affect the trajectories of other people's futures – it could even be as small as choosing to buy a coffee or driving to the supermarket.[11] Perhaps you slightly delay somebody waiting in the café queue so they don't meet their life-partner, or perhaps your drive subtly shifts the traffic so that a car crash is avoided. It seems far-fetched, but the average life is approximately 30,000 days, so the odds are that at least one of the moments during your time on Earth could significantly change someone else's path. One day, your acts will mean that the conception of a child does or doesn't happen: a different sperm meets a different egg, and that child grows up to live a different set of interactions across their 30,000 days of life too – and so on.

So, while deep time may feel enormous, and you might sometimes feel inconsequential within its long trajectory, you are not. If anything, it's the opposite. You exist because a chain of decisions made by millions of ancestors all went your way, and so too will your own choices ripple ahead centuries. Burke's partnership across the generations therefore extends far wider and deeper than first appears. Your life reaches into the deep past and future across thousands of years.

I wonder what Burke might have made of all this. Writing in the 1700s, he would not have known of deep time's full extent, nor the evolutionary and genetic insights that demonstrate quite how far one's ties extend across the ages. He wrote about generational partnerships in 1790, only a couple of years after geologist James Hutton's Unconformity had made his companions' minds grow giddy at the abyss of time, and two decades before Charles Darwin was even born.

However, Burke did give us a lens through which to approach long-term time that is just as relevant today – more than one, in fact.

His ethical call to consider our debts to the past and our duties to posterity has, more recently, been framed as 'Burkean longtermism' – a specific type of long-term thinking that differs from a purely future-oriented long view.[12] In short, Burkean longtermism is the belief that we ought to think more deeply about the role we play as we pass the temporal baton between our ancestors and descendants. It acknowledges that there is a longer trajectory of time, but it is person-centred, asking us each to consider more deeply what are we inheriting and what are we leaving behind.

The Burkean-inspired approach to the long term looks backward as well as forward, embracing the wisdom and learnings of forebears just as much as it encourages concern for descendants. As Burke wrote: 'People will not look forward to posterity, who never look backward to their ancestors.'[13]

As we'll see in the next chapter, Burke is far from the only person to have come to this conclusion, and he probably wasn't the first either: the idea of reciprocity between generations has emerged independently in various cultures, practices and faiths around the world. But nonetheless, I find his simple instruction to consider one's role in a long-term partnership to be a powerful framing, and I often find myself returning to it. It's easy to assume that taking the long view simply means looking ahead to the future. This is an important part of it, but I believe that a richer, broader perspective comes from projecting the mind across the entire arc of time.

This is not the only guiding principle we might draw from Burke's writing. Earlier in his life, he provided another lens that I believe could enhance our temporal perspectives: not so much an ethical call, rather a form of words for an emotional feeling. Burke may not have been aware of the billion-year timescales of the Earth, but when he was a young man, he wrote about the awe-inspiring magnitude of the *sublime*, and found evocative words for how it feels to encounter it.

He called it 'a delightful horror'.[14]

When Burke and his contemporaries were thinking of the sublime they tended to have the physical scale of nature in mind – Alpine mountains or vast oceans that dwarf the individual. But I believe the sublime is a concept that deserves rediscovery, and can help us understand the experience of facing temporal enormity too.

As Burke wrote: 'There are scarce any things which can become the objects of our senses, that are really and in their own nature infinite. But the eye not being able to perceive the bounds of many things, they seem to be infinite, and they produce the same effects as if they were really so . . . the imagination meets no check which may hinder its extending them at pleasure.'

The sublime defies the mind's attempts to hold it all in view. Encountering it, the poet Samuel Taylor Coleridge remarked, leads to 'a suspension of the powers of comparison'.[15] And as such, one feels a contradiction of emotions. As the philosopher Immanuel Kant put it, the sublime provokes a sense of powerlessness, akin to when one encounters 'threatening cliffs, thunder clouds towering up into the heavens . . . volcanoes with their all-destroying violence, hurricanes with the devastation they leave behind, the boundless ocean set into a rage, a lofty waterfall on a mighty river'.[16]

But crucially, there is also wonder – and even pleasure – to be found in sublime encounters. 'The passion caused by the great and the sublime in nature, when those causes operate most powerfully, is astonishment,' Burke wrote. Kant agreed, describing the sublime's power to 'elevate the strength of the soul'. Its immensity may provoke a daunting sense of smallness, but it can also be enriching, aesthetically and spiritually. When we gaze up at a giant mountain like Mont Blanc or consider a painting of a sea storm by J. M. W. Turner, these vistas are not only simply beautiful, but something more.

Later, the poet William Wordsworth would write:

> Of aspect more sublime; that blessed mood,
> In which the burthen of the mystery,
> In which the heavy and the weary weight
> Of all this unintelligible world,
> Is lighten'd[17]

Deep time as a 'delightful horror', then, can be uplifting as well as daunting.

The long view may often ask difficult questions of us: to imagine scales seemingly beyond our mental faculties, to embrace a grave duty to posterity and to accept the brevity of one's life within an overwhelming temporal span. But it starts by agreeing to a simple partnership, between those that came before you and those that will follow. And along the way, it may even bring an elevation of the soul, a lightening of this unintelligible world, and one or two moments of delight.

In the coming chapters, we'll discover more of the benefits that can be derived from embracing such long-mindedness. But first, a more personal story, about what the long view means to me, why I find it so valuable and how my temporal perspective has evolved.

The geology of loss

When I was nineteen years old, my father asked me to start collecting rocks for him. Back then, at the tail end of my teenage years, my view was still focused on the more immediate distractions of youth, so I didn't fully understand why. But it was easy enough to do as he asked: I was studying geology at university, and so everywhere I travelled, I would scan the ground for unusual stones.

At the lip of a fjord in New Zealand, where the mountains dropped into the water without pause, I picked up a piece of granite. On the flanks of the volcano Stromboli, which erupts twice an hour, I pocketed a bomb of lava basalt, with its gas bubbles frozen in place. Above a ski village in the Alps, a solitary day mapping outcrops was brightened by the prize of an ammonite, its fossil shell coiled like a ram's horn within a newly calved block of limestone. And as rain fell during a field trip in Scotland, I spotted the dark and pale stripes of one of the oldest objects you can hold in your hands: a 2.8-billion-year-old Lewisian gneiss.

My dad kept the rocks on a window-shelf at his office: stratigraphic souvenirs of the places that his son had visited. Yet I wouldn't know their significance until years later, when I would find them all over again, in a different place.

In 2010, my sister's first child was born, to my father's delight. It seemed like it would be a year marked by the creation of new generational ties – what I didn't know was that it would also be one where they would be broken. Around that time, Dad was in his early sixties and had decided to retire early so that he could spend his days fishing and birdwatching. He was new to both: he couldn't catch a salmon in a fish farm, and he mainly went to bird hides to chat with fellow twitchers, but we were all happy he was focused on his passions, rather than falling asleep in his armchair after work.

On 16 June, my phone rang as I was riding the lift up to my office. It was my mother, and it was an unusual time of day for her to call. In the confines of the lift, two women from another floor were laughing in a separate conversation, so at first I was unable to hear what she was saying. The lift doors opened and the women left, leaving me alone.

Shock does a strange thing to time. All I remember are semi-frozen moments: my finger on the lift button, walking to my boss's desk to tell him I had to leave, then an hour later, a half-empty train carriage on the route back home in the middle of the afternoon. Out of the window, London brick. Green pasture. Blurred stations. The world was no longer firm, with different tracks of my life sliding away from synchrony. I was slowing down, the landscape was speeding by, my dad had stopped.

On days like these, the perspective of deep time feels *untimely*. Amid an emergency, the past and future recede from view, until only the moment exists. This is as it should be. Otherwise, we could not be present for those we love.

I'm fortunate to have felt these temporal disturbances only a few times in my life: rare crises that have punctuated an otherwise happy and stable existence. One thing they taught me is that seeking the long view is a privilege afforded to those whose lives and surroundings are secure. For many people, one trauma follows another, so the past and future must feel continually unreachable. I'm aware that asking everyone to embrace deeper timescales, in exactly the same way, would be to ignore their circumstances.

That said, what I do know is that reaching for the long view in the wake of difficult experiences has provided insights about myself – and even forms of solace and purpose – that I may not have found otherwise. It would provide perspective in the wake of my dad's death. Following my daughter's stressful birth, it would lead to the discovery of Burke's generational partnerships. And more recently it helped once again. One day while I was writing this book, I lost my sense of time completely – and along with it, a future for my family that I thought I knew. I couldn't see how it'd ever return, but gradually it did, in a new form.

On a Monday morning in March 2021, my wife, Kristina, woke me before sunrise to tell me something felt wrong. She was thirty-five weeks pregnant, and that weekend we had been making final preparations for the birth. I had cleared a kitchen cupboard to make way for a row of bottles, the crib sat in the corner of the bedroom, and our hospital bag was packed. Sorting through our son's sleepsuits, we had tried to imagine his weight. We had decided to call him Jonah.

Worried about how she felt, Kristina called the twenty-four-hour triage midwife. Shortly after, we threw on some clothes, lifted our sleepy eight-year-old from her bed, and drove in the dark to the hospital.

During the Covid-19 pandemic, children weren't allowed in the maternity ward, so my daughter and I waited in the car park, while Kristina was alone inside. She had to phone me with the news.

When you're expecting a baby, you often imagine the paths ahead. Picturing this future, a word that comes to mind is 'potential', because when a new life begins, everything is open. By the time you're an adult, you've made many choices that have led you down certain tracks – sometimes choosing one branch has meant closing down another. But when you bring a child into the world, their own life has so much unspent potential that it can't help but add to your own, and it's wonderful.

If a baby represents a gain of potential in its purest form, so it follows that a stillbirth is the denial of possibility at its cruellest.

The moment I learned Jonah had died untethered me from a trajectory that I thought I knew. Or perhaps a better word would be a 'dislocation', because in that moment time fell out of joint. The rupture separated my unwitting past identity – the one who hadn't heard the words – from my experience in the moment. And as for my future self? Well, that was gone too. There was no expectation of potential any more, only a liminal present.

It would be dishonest of me to claim I was thinking about the long view in the middle of this experience. There was nothing that a reflection on deep time had to offer during an event like this. All I can do is describe how my wife and I began to navigate it together.

A few hours later, I held Kristina's hand as she lay in the operating theatre. The surgeons, nurses and anaesthetists moved around us with calm purpose. I'd been here before, eight years ago, for the birth of our daughter, Grace. That was a crisis of a different kind: back then, we had been rushed to surgery because Grace was distressed and an infection was taking hold. That time, I was mainly shocked into silence, unable to say much of value to my wife as she lay there. The anaesthetist guided us through it.

With that experience in mind, I had expected to be able to do nothing but focus on the operation in the present. But this time, Kristina and I managed to briefly step away from the awfulness of the moment. We were both frightened, but instinctively we shared memories from the past – the US road trips before we married or the first holidays to Europe's mountains with our daughter. And to my surprise, we also talked about the months that lay ahead: the close friends and family we'd soon see who had been separated from us by the pandemic, and the places we planned to go.

On one of the worst days of our lives, what helped was expressing gratitude for what we had known and seen, and thinking about the hopes we had for the future. That effort to mentally time-travel outside the present together was, at that particular moment, enough.

One day, a few weeks after Jonah died, I was walking with my daughter on the way to school. Our route takes us through a church graveyard. That spring, it was dotted with bluebells, snowdrops and long grass. The gardeners had been letting it overgrow to foster a wildlife reserve, letting the flowers grow wherever they emerged.

At the centre, I noticed a wide tree, with a split trunk that parts and rejoins around a hole, before splaying out above the gravestones:

The branches of the tree in south-west London.

Looking at its shape that day, I saw its branches as paths. If the trunk was the known present, then each of its splaying arms was one of myriad future possibilities. Walking on, through that landscape of bereavement and wild blooming, I began to feel a bit more intact. It was a small reminder that there is always potential.

It will always hurt to know that Jonah will never live to see this world. Here the long view feels painful: when I look to the decades ahead, I will always see an absence; the life he could have had.

But now every time I walk past the tree and its interarborating limbs, I see a truth about time that has guided me ever since: the past may be singular, but the future is always plural. And through that knowledge comes a sense of purpose. It shapes how I see the trajectories that lie ahead, and it shows me that, while I am alive, I still have a role to play in the partnership across generations.

Later, as my temporal perspective continued to recover and evolve, I would find other sources of solace within the long view – specifically, through the artefacts of deep time.

One afternoon, Kristina, Grace and I visited a pebble beach on England's south coast. It was chilly so we didn't linger, but as we sat, I was reminded of a memory and a ritual that had brought me some comfort after my father's death more than a decade ago.

A few years after we had lost Dad, I had returned to my mother's garden in Manchester and found something he had left behind. The moisture from the mossy grass and brown leaves seeped into my shoes. I came to a flower bed under a giant Leylandii tree. Not much can grow beneath these coniferous giants; the soil is starved of nutrients.

I spotted some of the rocks that I had collected for him, sitting on the soil, and was reminded of the promise I had made to him as a teenager. He must have moved them to the garden after he retired. Years had passed since I had last held them. I brushed off the dirt, and showed them to my mother. We talked about the father and husband we remembered. Later, she placed the rocks in a row on the wall outside her kitchen window, so she could see them every day.

The rocks on my mother's garden wall.

So, as I sat with my family on the beach, thinking about Jonah, I picked up a weathered pebble and took it home. I try to find a new one every time I visit the ocean, and have started doing the same again for my dad, whenever I travel to an exotic place.

In scientific terms, I know each rock has a different density, mineralogy and composition. I could label it as igneous, sedimentary or metamorphic if I chose. But when I pick one up it acquires a property that defies measurement.

Following this ritual, I am reminded of a mental practice that the geologist Marcia Bjornerud calls 'timefulness'. She defines it as the 'acute consciousness of how the world is made by – indeed, made of – time'. To be timeful is to feel an acceptance and reverence about the aeons that each of our lives sit within. And through it, she says, catharsis can be found. 'I am comforted by the knowledge that we live on a very old, durable planet, not an immature, untested, and possibly fragile one. And my daily experience as an earthling is enriched by an awareness of the lingering presence of so many previous versions and denizens of this place.'[18]

If I'm honest, I still struggle to reconcile that the sum of one's experiences – of clocks and calendars, of generations entwined, of joy, tenderness and grief – will soon disappear and become a mere flash within an unfathomable geological chronology. When I think of the indifference of deep time to the lives of my father and son, it can sometimes be overwhelming. But picking up a rock is my way of bridging those concepts in my mind – and not only that, it brings me a sense of timeful comfort. Collecting something formed long ago in the past, and applying an emotional weight to it in the present, is one of the best ways I know to pass forward what it is to be human into a deeper future. It is what the long view means to me.

So, when I look back at my own path to long-mindedness, leading up to the temporal perspective I hold today, I can see just how much it has fluctuated – and evolved. My interest in deep time began when I studied geology as a student, and I began thinking about the possibility of writing this book almost a decade ago. But along the way it has hardly been a story of steady temporal expansion. Like anyone, I have made many short-termist decisions in my life, and there have been moments when I have lost – and rebuilt – my sense of time completely.

What I have learned, though, is that the long view is much more than simply the ability to peer across deeper timescales. Sometimes, it is a perspective that allows one to transcend the stresses of the moment, and step into other times. Other times, it offers guidance in periods of uncertainty. And often, it provides principles for navigating the world. But most of all, it is a way of seeing more clearly what matters most within the here and now.

Over the years, I have also come to realise that there may be moral perils if you choose to immerse yourself fully in long timescales: a risk of becoming detached from the injustices and suffering of the present. The writer Robert Macfarlane expresses this well: 'There is a dangerous comfort to be drawn from deep time. An ethical lotus-eating beckons. What does our behaviour matter, when *Homo sapiens* will have disappeared from the Earth in the blink of a geological eye? Viewed from the perspective of a desert or an ocean, human morality looks absurd – crushed to irrelevance.'

We should reject this temptation. Instead, Macfarlane writes, we might see 'deep time as a radical perspective, provoking us to action not apathy. For to think in deep time can be a means not of escaping our troubled present, but rather of re-imagining it.'[19]

I will never directly experience the temporal perspective of von Baer's Millennium Man, but that's OK. In fact, if I did only see the world through his eyes, it would be a life cursed to miss so much. As von Baer wrote: 'We would not be able to register the change from day to night. We would not even register the Sun, but perceive it like a glowing piece of coal spinning around and around in a bright ring, we would see it only as an illuminated arc in the sky.'[20] Worst of all, the Millennium Man would have no idea of what the living world was like; all of humanity and nature would be far too transient to perceive. He'd have no concept of how anything moves, breathes or behaves – all he could do is make inferences from the long-dead fossils left behind, guessing at the content of rich, detailed experiences he'd never know.

We each have access to something better than this: a vantage point situated in the present – rooted in what it means to be human – that allows us glimpses of longer-term time if we dare to seek them out.

So, with this foundation established, let's now explore long-mindedness in all its forms and varieties. As we'll see, there are many manifestations of the long view, so let's start with some of the oldest: the temporal lenses of religion, ritual and tradition, taking in the world's longest-burning fires, an ever-renewing shrine made of wood and thatch, and more.

8

Timeviews:
Lessons from Faith, Ritual and Tradition

'Let us be such as help the life of the future.'

ZOROASTER

'Beginning is easy, continuing is hard.'

JAPANESE PROVERB

If you live in an industrialised rich country, it can be difficult to imagine that time could feel any different.

However, the *industrial timeview* that dominates much of modern life is a relatively recent development in human history. The proliferation of clocks, and their link to employment, meant that around 200–250 years ago, a certain form of precise time became far more ubiquitous, synchronising much of the world to the same chronology.

As the historian E. P. Thompson observed in his classic 1967 paper 'Time, Work-Discipline, and Industrial Capitalism', it wasn't always the case that a majority of people worked nine to five, clocking in and out simultaneously.[1] 'Within the general demands of the week's or fortnight's tasks . . . the working day might be lengthened or shortened,' he wrote. 'The work pattern was one of alternate bouts of intense labour and of idleness, wherever men were in control of their own working lives.'

The imposition of an industrial timeview, however, came coupled with specific values: a time-*ethic*. The increase in industries that used clocks to discipline and regulate their workforces was accompanied by a strain of moralism that would blame sloth and idleness for poverty, and praise hard work and diligence. 'Throughout the nineteenth century the propaganda of timethrift continued to be directed at the working people, the rhetoric becoming more debased, the apostrophes to eternity becoming more shop-soiled, the homilies more mean and banal.'

Industry made time into a resource to be mined and converted into profit; a transformation perhaps best evoked in 1748 by Benjamin Franklin in his short article 'Advice to a Young Tradesman', which scolded those who might use their time for idleness or leisure:

> Remember that time is money . . . the way to wealth, if you desire it, is as plain as the way to market. It depends chiefly on two words, industry and frugality; i.e. waste neither time nor money, but make the best use of both.[2]

Franklin wasn't the first to link time with wealth, but his words speak to the mood of his era. The vigorous capitalist emphasis on the financial value of time also meant that the future would eventually come to be targeted in the same way: as a space to be exploited for present-day gains, laying the ground for the blinkered, short-termist age to come.

One might assume that all other cultures had no choice but to be pulled along with the industrial timeview. But this would only be partly true. Over the centuries, many have developed their own temporal worldviews, which continue to transcend the pressures of capitalism. Both inside and outside the industrialised world, they have defined different manifestations of the long view, through faith, community and tradition. And each carries its own associated time-ethic, of stewardship, reciprocity and connection.

While it's not possible to rewind the clock on industrialisation, that doesn't mean you can't loosen industry's grip on the clock.

So, what are these alternative *timeviews*, and what might be learned from them?

It's tempting to think that to shape the future, and pass something forward to your children, you have to create an heirloom which will endure. It's why people through the ages have built libraries, erected monuments, or fashioned exquisite objects. If you consider the Pyramids of Egypt, the Taj Mahal, or Stonehenge – or precious antiques like a Ming vase, Renaissance painting or Fabergé egg – this is an approach that can sometimes work.

I call this attempt to make enduring heirlooms the *Patek Philippe strategy*. Why? For decades, the watchmaker has run advertisements featuring a parent and child with the slogan: 'You never actually own a Patek Philippe. You merely look after it for the next generation.'

In practice, however, it is very difficult to create something that will be venerated and preserved over long-term time. Create a fixed legacy that is designed to be passed forward, and it will always be vulnerable to future changes. One day your descendant loses the watch, closes the library, or abandons the monument and the legacy is gone forever.

And if this is true for physical objects, it's even more so for *ideas*. Let's say you want to pass on a piece of knowledge or experience to future generations: it could be a belief, a value, or a warning. You have lived through a pandemic and seen the awful human toll. You saw the mistakes that were made and you wish to ensure they will not happen again. How do you do that? Write down a warning to the future and hope that it is read? Build a statue that you hope people will visit?

Like a building or object that needs maintenance over time, so too does an idea or piece of knowledge. Most information never fully dies – it may lie in an obscure history book, informal family history, or on a statue inscription – but that doesn't mean ideas can't gradually fade from collective memory until they are no longer relevant, or observed.

There are, however, ways to ensure ideas get passed on – and you can find the most effective approaches in religions. While most faiths have their physical artefacts – places of worship, relics and sacred objects – that is not what makes a belief system endure. It's something different. Unlike the mindset of the industrial timeview, the emphasis here is on actions and behaviours that foster longevity – both of the faith itself and the community practices, moral teachings and generational ties attached to it. I call this the *continuity timeview.* And it's one of humankind's oldest manifestations of long-mindedness.

To better understand this perspective, let's explore a couple of examples – from an everlasting fire in Iran to a Japanese temple that is methodically rebuilt every twenty years.

The fire that never goes out

At some point in the late fifth century, as the Western Roman empire fell, a group of Zoroastrian priests in Iran's Fars Province lit a very special fire. As the days passed, the flame was kept burning. Years became decades, and decades became centuries, with the fire moving between various locations, until it eventually ended up in the Yazd, a desert city around 600km (373 miles) south-east of Tehran. In 1934, a new temple was built there to house it, where it continues to burn to this day. It's one of only nine in the world – a flame that has been kept alive for more than 1,500 years.

The Yazd temple today is situated on a busy street with cafés, clothing stores and a tourist information centre. Once you are inside the gate, however, the world outside fades into the background. Visitors encounter a peaceful garden, containing a round pool of water lined with benches and conical trees. Beyond that is a light-coloured, one-storey brick building, with a portico topped by the Zoroastrian 'Faravahar' symbol: a bird's wings outstretched like an aeroplane viewed from above, with a holy male figure for a head.

Inside the building, the everlasting fire burns within a goblet. Several times a day, priests wearing all white tend the flames with a mixture of long-burning hardwood and sweet-scented softwood. Non-Zoroastrians are not allowed to go close, but visitors can view the chamber from the entrance hall. Looking at the fire through a tinted glass window, you can see the faint reflection of tourists peering in with their cameras, attempting to capture an image that will no doubt have faded or digitally decayed long before the flame goes out.

Zoroastrianism is one of the world's oldest faiths, and was founded approximately 3,500 years ago. It is based on the teachings of the Iranian prophet Zarathustra (also known as Zoroaster). In the Yazd fire temple, he is depicted in a painting with a bushy beard and long hair, a halo behind his head, carrying a staff and holding up a single finger, his eyes gazing upward.

With believers concentrated mainly in Iran and India, Zoroastrianism is much smaller than the major global religions: between 100,000–200,000 followers by some estimates.[3] But over the centuries, Zoroastrian practices and writings have significantly influenced other faiths, as well as intersecting with the politics of states and empires. It gave Christianity the three wise men who attended the birth of Jesus – scholars reckon they were Zoroastrian priests – and supposedly helped to inspire Judaism's theology of the afterlife, with the idea that what you do on Earth affects your fate after you die. Also, in the Book of Isaiah, the description of light as 'good' and darkness as 'evil' shares uncanny parallels with Zoroastrian verses attributable to Zarathustra, hinting at a dialogue between the two religions thousands of years ago.

Zoroastrians have a particularly strong relationship with fire, which they see as a focus for ritual and contemplation. The ancient flames they tend are called Atash Bahrams, which means 'victorious fire'. Most are in India, which has a minority Zoroastrian community known as Parsis, whose ancestors fled Islamic persecution in Iran in the seventh century.

The fires are not worshipped, but when standing nearby, believers feel they are in the presence of the deity Ahura Mazdā. The flame can be symbolic of various things, expressing inspiration, compassion, truth, devotion, as well as continuity and change.[4]

Atash Bahram fires are *extraordinarily* difficult to start, which explains why there are so few of them. The oldest fire in India, for example, has stayed burning for more than 1,000 years in a village called Udvada, north of Mumbai. To start it, Zoroastrian priests had to walk back to Iran to fetch a collection of sacred items called the *alat* – such as holy ash, a ring and the hair of a bull. En route they had to hide to avoid enemy armies and could not cross any rivers or seas, because fire and water cannot mix. It then took 14,000 hours of ritual. But here's where it got really difficult: an Atash Bahram must be made by combining sixteen different fires, taken from the homes of various professions such as a bricklayer, baker, warrior and artisan, plus the fire of a burning corpse and the fire of lightning. The latter fire is particularly difficult to source, because two Zoroastrians have to witness the lightning, and within a rainy storm hope that the strike sets something alight.[5]

It is of course impossible to verify if the ancient fires have ever fizzled out once or twice. One can imagine that the chain has been disrupted by war, disease or natural disaster – and across 1,500 years there must have been many close calls. But the tending of the Atash Bahram flames is nonetheless one of the world's longest-term commitments to a single act. And remarkably, it has endured through the medium of one of the world's most ephemeral substances: a flame.

So, what elements of the Zoroastrian faith led to this longevity, apart from pious dedication? Before we answer that question, let's look at one more example of the continuity timeview. Here long-mindedness is carried within another transient medium: the wood and thatch of a temple, embedded in a forest in Japan.

The ever-renewing shrine

One day in the seventh century, followers of the Shinto faith in Ise, Japan, began to rebuild their grand Jingu shrine – and it wouldn't be the last time. They have methodically and deliberately reconstructed it sixty-six times, in a practice that has lasted for 1,300 years.

Every two decades, priests of the Shinto faith perform a ritual in ceremonial robes to mark the completion of the new building, carrying boxed treasures between the old temple and the new. The renewal system is called Shikinen Sengu and involves the passing of tradition and craft from one generation to the next. Apprentices learn from master builders, woodworkers from their seniors and priests from their elders.

To put this remarkable act of continuity in context, it helps to know a bit more about Shinto. The indigenous faith of Japan, Shinto is at least 8,000 years old as a religion, with an emphasis on the values of harmony, collectivism and cooperation. (Its principles have arguably spilled over to the nation's longer-minded business culture too, as we learned in Chapter 2.) Its followers worship kami: spirits that are entwined with the natural world, representing oceans, mountains, wind or rain. Ancestors are also venerated as ancient family guardians, and individuals who have made a significant contribution to society may too be remembered as kami. Each spirit supposedly plays a role in the ordering of the world, meeting to discuss it. For Shinto believers, appeasing the kami is an important way of pacifying the harsher aspects of nature, such as drought, disease, famine, tsunamis and storms.

While there is no doctrine or founder, nor a single omnipotent deity, the Shinto faith has origin stories that have been passed through generations. One of the more important, particularly relevant to the shrine at Jingu, is the tale of Ama-no-Iwato, the Celestial Cave.[6] It goes like this.

At the beginning of Earth's history, a deity couple called Izanagi and Izanami gave birth to the Japanese islands and other kami. Their descendants Amaterasu-Omikami lit the heavens, Tsukiyomi-no-kami represented the Moon and night, and Susano'o-no-kami was associated with the sea and storms. Susano'o was impetuous and idle, however, and abandoned the oceans to join Amaterasu in the celestial plain. His troublemaking behaviour there forced the Sun goddess to hide in a cave, and the light went from the world. The kami met to discuss how to fix the problem, and decided to use a jewel and a mirror to decorate a tree near the celestial cave. They then held a joyful ritual with dancing, and managed to coax Amaterasu back out. Susano'o was repentant and descended to Earth, where he slayed an eight-headed monster. Discovering a sword in the monster's tail, he dedicated it to Amaterasu.

Amaterasu was originally worshipped at the Imperial Palace in Tokyo, but following an epidemic around 1,300 years ago, officials decided to move her symbol – the sacred mirror – to a new location where she could be worshipped with less disruption, and more respectfully. Japan's princess searched the country and after a revelation by the banks of the Isuzugawa River, she eventually decided on Jingu, in Ise.

There are around 80,000 shrines around Japan, known as jinja, but the one at Ise is particularly important, and marks the end of a pilgrimage route walked by foot. There, Shinto followers and priests hold regular ceremonies called matsuri, often tied to the calendar, such as thanks for crops, prayers for future harvest or to honour the Imperial family.

The Sengu ceremony to inaugurate the new grand shrine is particularly rare and special, however, because it is held only once every twenty years. The tradition has continued for thirteen centuries, only occasionally being suspended, in times of war.

The preparation leading up to it lasts eight years, beginning with the felling of cypress trees from around the region. Many of the logs are cut in the mountains and floated down-river manually. In the past, the wood-workers felled cypress trees from a forest nearby, called the Misoma-yama. But over the years it became increasingly hard to source the 10,000 logs required. During the Edo period between the seventeenth and nineteenth centuries, millions of pilgrims came to Ise to worship, but they used a large amount of the trees from the local forest for firewood. So, around ninety years ago, shrine officials began to plant new trees for future generations to use. A quarter came from this source for the last ceremony.[7]

Once the Shinto woodworkers have collected and assembled the shrine's timber, the first four years are spent curing and preparing: soaking the wood in a pond to extract oil, acclimatising in the air, then sawing.[8] Finally, the beams, pillars and joints are placed on decorated carts and pulled by locals to the site, accompanied by singing youths.

Gradually, the main shrine – the palace of Amaterasu – is replicated, beam-by-beam, until there are two buildings side by side. The design is similar to a traditional grain warehouse, with a raised floor, pillars and thatched roof.

*The previous (behind) and new Naiku complex of Ise Grand Shrine
in September 2013.*

Intriguingly, over longer-term time the building design is more fluid
than first appears. It's tempting to assume that it follows a strict blueprint
created by the original architect, but today's version has subtle differences
to the very first one. For example, gold-copper fittings were introduced
during waves of technological change and cultural immigration from
China hundreds of years ago.[9] The shrine would never be replaced with
concrete or some other drastic departure from tradition, but crucially, it
is a living building, where each generation has an opportunity to adapt
it, should they want or need to.

When the new shrine is ready, a ceremony is held, attended by a
member of the royal family, to move Amaterasu's sacred mirror to its
new home. It's not really intended for public entertainment, so only
a few hundred onlookers attend. Alexander Rose of the Long Now

Foundation, who was there during the last ceremony in 2013, described what he saw:[10]

> A series of chests were roped off in the courtyard area. Some of these were plain wood and some of them were lacquered. These chests contained the temple 'treasures' that are moved from the old temple to the new. Some are re-created every twenty years by the greatest craftspeople in Japan, some have been moved from temple to temple for fourteen centuries, and some are totally secret to all but the priests. The treasures are what the Kami spirits follow from one temple to the next as they are rebuilt. So the Shinto priests move the treasures when the new temple is ready, and the Kami spirits move sometime in the night to follow them in to their new home.
>
> Without fanfare, the princess of Japan led a march of hundreds of Ise priests down the path that we had just walked, and they all lined up in rows next to the chests. After a ceremony with nearly 30 minutes of bowing, the chests were carried into the sanctuary and placed into the new shrine . . . All very calm, very simple, and without any hurrah.

A print from 1849, depicting priests carrying treasures between the temples.

Following the ceremony, the old shrine is dismantled. The ageing cypress timber is not discarded, rather refashioned to make arches elsewhere on the site, frames for screen prints and to build other shrines elsewhere.

Then, around twelve years later, preparation for the new shrine begins once again, culminating another eight years after that. So, if you happen to be passing through Ise sometime between 2025 and 2033, you might get to see the next shrine coming together – the sixty-seventh time it will have been built.

Ritual bonds

The everlasting flames of Zoroastrianism and the sacred temples of Ise show that it's not necessary to leave behind something designed to last forever if you want to bridge across the long term. While both Shinto and Zoroastrianism certainly have their precious treasures, such as Amaterasu's sacred mirror or the *alat* used to start an Atash Bahram, arguably these faiths' most valuable heirlooms are instead their community practices and habits. It is these that define the continuity temporal perspective.

Like so many faiths and cultures, Shinto and Zoroastrianism emphasise that there is a bond between generations. Through a shared act, the shrine rebuilding in Japan ensures that each generation is trained up within a community with expert skills, from woodworking to sacred duties. And by focusing attention on a fire that must be tended, the Zoroastrians pass a sacred responsibility forward.[11]

What makes this so powerful is that along the way, these individuals personally benefit while also being inculcated in a belief system that can be passed forward to their children. Unlike the Patek Philippe strategy, where the only incentive is to be the custodian of a fixed legacy, each new generation therefore experiences the same personal rewards and status as the previous one.

As the Chinese philosopher and politician Guan Zhong is believed to have remarked back in the eighth century BC:

To plan one year ahead, plant grains.
To plan ten years ahead, plant trees.
To plan a lifespan ahead, plant people.

But this is not the only long-minded lesson we might draw from the continuity timeview. Another crucial way that faiths like Zoroastrianism and Shinto – or indeed any successful religion – pass their ideas across time is via the power of *ritual*.

The performance of ritual can be traced into human prehistory. Back then, organised religion didn't exist and people lived in small communities. So, they were perhaps more likely to have followed the kind of infrequent-but-traumatic rituals seen in forager groups today. In the Sepik region of Papua New Guinea, for instance, men participate in ceremonies where they carve their skin to resemble a crocodile's.[12] When everyone knows each other and cohesion is vital for survival, such painful rituals allow individuals to display a tangible willingness to commit to the needs of the group.

But as societies grew larger, the more routine community-building rituals of faith came into their own, such as prayer, music, tending flames, ceremonies and more. According to the anthropologist Harvey Whitehouse at the University of Oxford, rituals helped to foster the trust, cooperation and cohesion that enabled civilisations to flourish: a social glue that bound people together across space and time.

Rituals helped to spread the idea of what a 'good' citizen should be, gluing together heterogenous societies. Every time a prayer was recited or a ceremony performed, it signalled a commitment to shared moral beliefs and collective goals among disparate people. When individuals did not know each other, separated by distance or time, it was the performance of the ritual that reminded them that they are connected, fostering pro-social behaviour, trust and cooperation. As the Islamic scholar Ibn Khaldun observed in the fourteenth century, rituals fostered '*asabiyah*', which in Arabic roughly means 'social cohesion', transporting solidarity beyond direct kinship to a national scale.[13]

Over time, ritual practices became ever-more embedded in the major organised religions – Christianity, Islam, Hinduism, Buddhism, Sikhism, Judaism. They are all different in detail, but have much in common. Many involve synchrony or display, such as the Islamic call to prayer or the Christian singing of hymns. Food makes a regular appearance, such as in Catholic Communion, or the Buddhist preparation of meals to feed hungry ghosts (a neglected spirit or ancestor). Fire or burning incense is also seen across countries and faiths – the lighting of candles to mark the start of the shabbat, or the diya lamps during Diwali. And so is cleansing, such as the various procedures followed before entering temples, or the Hindu practice of bathing the body in holy rivers before festivals.

For the Zoroastrians, tending the fire is a ritual in itself, and the locus for regular ceremonies to mark occasions, called jashan, which involve implements such as fruits, nuts and wheat pudding in metallic trays placed on a white sheet with milk, wine and flowers, led by a priest called a zoatar, while another person looks after the fire: an atravakshi. And for the Shinto priests, there are three timescales of ritual. The Sengu twenty-year rebuilding is one cycle, but there are also daily or annual events, such as to give thanks for a good rice harvest (Kanname-sai), as well as special occasions to honour the Imperial family and Japan's general prosperity.

Plenty of rituals have no obvious reason to be performed in the specific way that they are, and one culture's ritual norm can raise eyebrows in another. But the detail does not matter. It's about the ideas they carry, and the community behaviours they help to foster.

As well as encouraging repetition and remembrance, these rituals are a way of forging a relationship with longer-term time, marking beginnings and endings, as well as a connection with ancestors. By following a specific procedure with tradition and piety, they allow an individual to feel part of a community across past, present and future in pursuit of something bigger than themselves. Rituals therefore are a human behaviour by which ideas can travel across decades and centuries.

If a non-believer or secular organisation were hoping to become more long-minded and create ideas that endure, they might do well to ask: what rituals and traditions bring *their* communities together?

Some rationally minded sceptics might be reluctant to participate in a spiritual practice, but not all rituals involve deities or worship. Nor do they have to be solemn or serious. One of my favourite long-minded examples is the once-a-century Mallard ceremony at Oxford University, the last one of which was held in 2001.

Once every 100 years at All Souls College, a group of eminent academics parade around the quad singing a song about ducks, while carrying a wooden mallard on a pole. The torchlit procession is led by one of their number, who takes on the title of 'Lord Mallard' (as far as I know, there has never been a Lady Mallard).[14]

It's a tradition with roots in the Late Middle Ages: supposedly a duck flew out of the college foundations as it was being built in 1437.[15] The chorus goes like this:

> Hough the bloud of King Edward,
> by ye bloud of King Edward,
> It was a swapping, swapping mallard!

The song is so old that it's unclear which King Edward is referred to, or why. In the 1600s, one account describes complaints after it 'was sung after a rude manner about . . . 2 or 3 in ye morning, which giving a great alarm to ye Oliverian soldiery then in Oxon they would have forced ye gate open to have appeased ye noise'. So, if one of your descendants is outside All Souls College late one evening in 2101, they might well hear it.

Parading a duck is one of the more unusual secular rituals, but there are many more everyday ones that aren't so flamboyant. Ask yourself what you have in common with your ancestors, and it's events like celebratory meals, playing games and celebrating festivals; eating your culture's food, marking marriages and births, or mourning deaths. These are the perennial signatures of the human condition. In a few hundred years' time, people will be using radically different technologies, but they will almost certainly still follow rituals – they are one of the most long-minded habits we have.

Of course, the continuity timeview is not the only alternative temporal perspective with roots in faith. Another is what we might call the *transcendental timeview*. We've touched on this a few times already, so I won't repeat the details, but one of its manifestations is the expectation of a looming end for human experience, via death (individuals) or the apocalypse (society) followed by an infinite existence in heaven. If there is a future, it is one that will shortly draw to a close, and after that time would be perennially the same, perhaps even consciousness too. Importantly, this view comes coupled with a time-ethic: encouraging compassion, devotion and kindness in this life, so that you may be rewarded in the next.

Not all of these transcendental timeviews involve apocalypses or ends though: some don't feature time at all. In the 2,000-year-old Chinese Daoist text the *Zhuangzi*, there are immeasurable non-times of nothingness, or cosmological 'primal chaos' without order or sequential linearity.[16]

Then there are transcendental timeviews that combine an intriguing mixture of cyclical and deep-time perspectives. These involve awesome chronologies similar to geological time, but rotating between destruction and rebirth. For example, Hindu and Buddhist cosmology talks of kalpas, a unit of time that defines the period between the creation and recreation of the Universe. The length varies, but in Hinduism it's defined as 4.32 billion years. There are even longer versions: in the *Bhagavad Gita* scripture, written more than 2,000 years ago, the lifespan of the Universe is calculated to more than three trillion years.

Over the centuries, religious scholars have deployed evocative metaphors to help highlight just how big a kalpa is. For example, some defined it as the time it would take to erode a mountain to nothing if you brushed it once a century with a silk cloth or an eagle's wing. Another describes the time it would take to fill a cube 26km (16 miles) wide and 26km (16 miles) high with mustard seeds, at a rate of one every 100 years. But my favourite involves an unlikely turtle: imagine throwing a wooden yoke (a collar for beasts of burden) into the ocean once a century. A kalpa is how long you'd have to wait before a one-eyed turtle happened to emerge at the surface exactly within the yoke's hole.[17] None of these metaphors is scientifically precise – mountains don't last billions of years and nor do turtle species – but that's not the point. The goal is to emphasise how short individual experience is within the extremely long timescales of a transcendental world.

While this timeview would seem to echo (and significantly predate) the perspective of geological deep time, its emphasis on rebirth might instead make it closer to some of the theories of physics. A few cosmologists propose that the Universe, over the extremely long term, is cyclical. So, there was a Big Bang, but it'll be bookended by a Big Crunch, where all of space–time collapses back on itself to restart anew. In this view, the Universe expands and contracts, forever.[18]

For those (including myself) whose temporal perspective is rooted in geological knowledge, it's humbling to reflect that there could be even deeper long views that transcend my own understanding. I operate on the assumption that time is linear and did not exist before the birth of the Universe . . . but what if future scientists were to discover beyond doubt that time has no beginning or end? What a sublime encounter that would be. Long-mindedness comes in many forms, and I have to continually remind myself that my own particular long view is built on knowledge and assumptions about the world that may well be incomplete.

Indigenous insights

Faith, of course, is not a prerequisite for long-mindedness and a different view of time. Plenty of other non-industrial cultures have also developed their own temporal perspectives – particularly indigenous groups.

E. P. Thompson pointed out that many non-Western cultures have continued to prioritise their lives and communities by aligning with their immediate surroundings, rather than by industrially synchronising via minutes and seconds. A farmer might work according to the light and a fisherman to the ocean, Thompson suggested, and in some cultures, like one he cited in Madagascar, 'time might be measured by "a rice-cooking" (about half an hour) or "the frying of a locust" (a moment)'.[19]

Collectively, this lens might be described as an *environmental timeview*, which embraces a temporal perspective more closely aligned with surroundings, including natural processes, creatures and cycles. Within the indigenous cultures in which it continues to be embraced, it is also often accompanied by a *generational timeview*, in which past, present and future are perceived through the lens of one's close relationships and responsibilities. (This latter perspective echoes the Burkean view of posterity that we encountered in the previous chapter, and also overlaps with the continuity timeview . . . but as we'll see, it extends a little deeper.)

Perhaps the most famous manifestation of these alternative temporal views within indigenous culture might be the Aboriginal Australian perspective of time, which is sometimes described as *everywhen*. It's difficult to conceptualise if you are used to thinking about the past, present and future as a line or a landscape, but some have attempted to describe it as perceiving time as a body of water that surrounds you. Within the water, the past and present are the same, and the future is not relevant. Aboriginal people are of course quite capable of understanding chronological time if required, but they embrace the alternative as a cultural worldview. The clock, in other words, is not the dominant controller of life.[20] Instead, an environmental and generational timeview is the guide, marked by the priorities of aligning with one's ecological surroundings, social relationships, ancestors and community.

Consider also the environmental and generational lens of the Idu Mishmi people, who live in the Dibang Valley of Arunachal Pradesh in north-east India.[21] Their home is a densely forested biodiversity hotspot that contains more than 550 species of birds. It is also home to a new population of endangered tigers, which conservationists credit the Idu with helping to foster. The Idu are animists, who believe that all natural things are spirits that share a common culture with humans. The tiger is part of the people's origin story, as told in songs, so killing them is forbidden. Where this relates to time is through rules called ena that emphasise ecosystem balance and generational baton-passing. As Sipa Melo, a spiritual leader with the Idu once put it: 'No matter how developed and successful you might become, if you don't save the forests, mountains, rivers and lakes now, no God will save you . . . Nothing will be left for our children and grandchildren.'

Such attitudes can be continuity-minded too. In the same way that ideas are passed forward through religions, so too can so-called 'Traditional Ecological Knowledge', which can cover everything from farming practices to controlled fire burns to animal behaviour. It is experience that takes a very long time to acquire, and is only built up by living within a place across generations – over hundreds or even thousands of years.[22]

The researchers Henry P. Huntington and Nikolai Mymrin once discovered just how insightful and valuable this long-acquired knowledge can be when they were attempting to study beluga whales in the Arctic. When tracking the behaviour of animals over longer-term time, there's a limit to what scientific instruments can reveal – unless you leave them in place for years or decades.

But by talking with indigenous communities in Alaska and in Chukotka, Russia, Huntington and Mymrin were able to document novel details about the timing, location and direction of belugas in the region, which scientific tools could not have accessed so readily. This included unique insights on annual migrations, as well as details of the whale behaviour, such as feeding and calving. 'People have relied on this detailed knowledge for their survival,' they wrote afterward. 'They have literally staked their lives on its accuracy and repeatability.'[23]

The pair also learned things that they didn't expect, or even thought to ask about. 'During one group interview session, we were suddenly talking about beavers instead of belugas. I wondered if I should try to steer the discussion back at least to the ocean, when one of the participants enlightened me about the connection. Beavers dam streams where salmon and other fish spawn. Since the beaver population is increasing, this may mean loss of spawning habitat, changing the fish populations that the belugas feed on. Therefore, the beavers' activity may affect the belugas.'

Meanwhile, the generational timeview embraced by many global and indigenous cultures also means that the value of intergenerational reciprocity is brought to the fore. There are myriad festivals that celebrate ancestors, for example. In Mexico and the Americas, there's the Día de los Muertos (Day of the Dead). In China, the Hungry Ghost Festival. And in Cambodia, the festival of Pchum Ben.

Then there's the Maori proverb '*Ka mua, ka muri*', which, as we touched on in Chapter 6, translates as 'walking backward into the future', emphasising how the learnings of past generations can provide a guide to what's ahead. Respect your ancestors, goes the wisdom, and they can help you in return. Maori communities also have a much-valued genealogical concept called whakapapa, which describes ancestry and lineage. It's conceptually different to the metaphor of a family 'tree', in that it is more often described as a narrative, told layer by layer, with each generation overlaid on another – perhaps more like a sedimentary rock that grows over time rather than branches extending into the past.

It's interesting to consider how this metaphor might encourage a different mode of thinking compared with that of a tree. A family tree places the individual at the centre – the trunk – with ever-finer branches reaching towards relatives into the past. Layers of rock, by contrast, place ancestors at the base, growing in size and scale over time. It accounts for both past and future, with the promise of accumulating layers lithifying into an intertemporal community.

But if you had to highlight where the generational timeview has been most prominently and influentially articulated, it would be among Native American culture, and the idea of 'seventh generation' stewardship.

This lens on time is thought to go back to the centuries-old Great Law of the Iroquois confederacy, which included the Mohawk, Oneida, Onondaga, Cayuga, Seneca and Tuscarora nations, living across the eastern US and Canada. While seven generations aren't mentioned directly in this constitution, that's perhaps not too surprising given that it is at least 500 years old – originating sometime between 1390 to 1500 – and was orally passed on until translators later wrote it down in English.

The English transcription of the Great Law does, however, contain a maxim calling on leaders to 'look and listen for the welfare of the whole people and have always in view not only the present but also the coming generations, even those whose faces are yet beneath the surface of the ground – the unborn of the future Nation'.[24]

In the present day, there are different interpretations about what the seventh generation principle means. It is most often seen as a call to do right by descendants living seven generations ahead: approximately 150–200 years from now. This forward-reaching view has been enthusiastically embraced by activists campaigning for the rights of future generations outside Native American culture, as well as the founders of the US consumer brand Seventh Generation, now owned by Unilever.

However, the now-deceased Native American scholar Vine Deloria Jr suggested that the seven generations needn't refer solely to the future unborn, and supposedly expressed annoyance at the romanticism attached to the concept. When he himself first came across the term in the speeches of past leaders, he had assumed it was a way of saying 'a long time' in a culture without precise timekeeping. But on closer inspection he believed that wasn't the case.

Instead, Deloria suggested the intention was to describe a *symmetrical* reach of generations across past and future – so, extending from great-grandparents through to great-grandchildren. 'Each person, we might say, is the fourth generation and looks back to three generations and forward to three,' he wrote in 1988. 'When the old chiefs spoke of the seventh generation they were basically saying that they wanted their great-grandsons, whom they hoped one day to seem to have the same rights and privileges as they themselves did. So instead of being a vague term for time, seven generations has a reality and precision within the family context as specific as any written contract ever drawn.'[25]

The symmetrical interpretation 'makes so much more sense on a human scale and does away with the destructive myth of mystical, all-seeing Natives', writes the academic David E. Wilkins, a citizen of the Lumbee Nation.[26] 'In truth, our peoples were visionary but not in a passive, new-age way. We actively tended our families and our clan-ties by holding the lives, memories, and hopes of all Seven Generations close. Each generation was responsible to teach, learn, and protect the three generations that had come before it, its own, and the next three. In this way, we maintained our communities for millennia.'

But no matter which interpretation is the most historically accurate, both have power. Forward-facing or symmetrical, the broader point is that a generational timeview like this can translate what can sometimes feel like an impersonal 'duty to posterity' into the language of family, close ties and community. It's sometimes difficult to extend empathy across deep tracts of time, but if you start with those nearest to you, it begins to get a little easier.

The industrialised, consumerised culture of the twenty-first century has come to utterly envelop the attitudes and moral values of those that live within it – and that includes how people think about time. However, as we've discovered from our tour of alternative religious and indigenous timeviews, there are many different temporal perspectives around the world. While each culture may have developed its own approach to the past, present and future, there's no reason why the ethical principles under-pinning these timeviews could not be more widely embraced: long-term stewardship, a focus on continuity, a closer connection with surroundings and intergenerational reciprocity. All are universally human aspirations, and therefore could be integrated into everyone's temporal perspective.

With this in mind, let's now move on to explore another timeview rooted in ethical principles: this one emerged only a few years ago, but it has already transformed how thousands of people see their moral respon-sibilities to the future. It's an approach called *longtermism*, and the story of its origins and implications is worth telling in detail, because if you haven't heard of it already, its accelerating growth as a social movement suggests you soon will.

9

Longtermism:
The Moral Case for Prioritising Future People

'Why should I care about future generations? What have they ever done for me?'

ANON[1]

'One point should perhaps be emphasised more particularly; it is assumed that we do not discount later enjoyments in comparison with earlier ones, a practice which is ethically indefensible and arises merely from the weakness of the imagination.'

FRANK RAMSEY[2]

Imagine a forest. Within it, there's a clearing where somebody, long ago, dug up the soil. Just beneath the surface, they left something dangerous.

Between the trees, a child is playing barefoot. Her toes press against the mossy floor, kicking up fallen leaves and the scent of the undergrowth as she runs. She is only a short distance from the clearing.

Now picture a similar scene in 100 years' time. There's another little girl, playing the same game in the same place, with the same bare feet. Some of the trees have fallen, and some have grown many more rings, but the clearing is still there.

One of the two children comes to the dangerous place and begins to run across. Suddenly, they cry out in pain. Hidden just below the grass there are hundreds of shards of glass.

Which would be worse: that the present-day child steps on the glass, or the child in 100 years' time?

This scenario was proposed by philosopher Derek Parfit as a way to challenge how we think about harms to people in the future. Presented in this way, it would be hard to argue that there was any moral difference between injuring either child: both are equally bad.

Yet in practice, our generation is leaving many harmful legacies for future generations. Not glass, but more malignant heirlooms, such as plastic fibres in the ocean, spent nuclear fuel rods, or a heated atmosphere. As we learned from Bruegel's painting of Icarus in Part II (see page 121), the further that people project their mind into the future, the more abstract and psychologically distant its inhabitants become. And this in turn shapes the priorities of politicians, the acts of organisations, and the day-to-day choices of individuals.

In recent years, however, an influential and well-funded movement of philosophers and philanthropists has emerged that is seeking to challenge and expand people's moral attitudes towards these psychologically distanced future people. It is a school of thought called *longtermism* – and in many respects, their beliefs are the most extreme example of long-mindedness you will encounter in this book.

Inspired by Parfit and his ideas, these longtermists believe that people in the future should count for far more in our present-day decisions – and not just those living in a few decades' time, but thousands or millions of years away. Whether it's the child today or the child tomorrow, the lives of both should be considered equally, even if one doesn't exist yet.

While longtermism might sound similar to 'long-term thinking' in name, it actually goes much further, by looking at the future through an ethical and mathematical lens that, if correct, might demand more of our generation than the majority are currently willing to do. It suggests that when we think about the future, we ought to consider the morality of how our actions today have an impact on humanity across time – but the argument is not primarily concerned with our everyday duties, or a religious sense of right or wrong, instead on calculations that weigh up the welfare of trillions of people.

You may find that you do not agree with everything that the long-termists believe – some of their views would seem to imply demanding acts and controversial conclusions – but their approach nonetheless offers a whole new lens for thinking about our roles and responsibilities within deep time. So, what exactly is longtermism, and what might we take from its ideas?

Parfit and the bottomless pit

If you had to trace back the philosophical roots of longtermism, then most paths lead to Derek Parfit. A philosopher at Oxford University, Parfit was hugely influential before he died in 2017. He was also long-minded in so many ways: in the subject of his work and the conclusions he drew, but also in the choice he made to tackle seemingly insurmountable problems during his career.

One way to understand Parfit is through his eccentricities. With a wild crop of white hair and a disregard for small talk, he became known for his unusual behaviours. 'Gradually a legend built up around him,' writes S. J. Beard at the University of Cambridge, who knew him later in his life. '"Derek only eats meals he can consume with one hand so he can read and eat at the same time." "Derek drinks instant coffee made with hot water from the tap, so he doesn't have to wait for the kettle to boil." "Derek always wears the same clothes, even in the St Petersburg winter, to spare him from having to think about what to put on in the morning." Unusually for such legends, this was all completely true.'[3]

However, most people influenced by Parfit know him best through his ideas. He cared about the quest for objective moral facts, even if they seemed out of reach. What marked him as different to many philosophers of his generation was that he was willing to take on difficult, unfashionable and potentially unsolvable problems about morality and well-being on the scale of humanity, which Beard describes as 'bottomless pits of doom'. But by being willing to take the time to explore these pits, he would gain treasures that no one else had found, says Beard, and identify questions that his successors would go on to explore decades later.

One of the most influential of those questions can be boiled down to this: what exactly do we owe to future people? Parfit believed that the lives of future generations mattered, and that we could do more to consider how our choices today affect them. With his analogy about the children stepping on the glass in the forest, he was making the point that we have moral obligations not to harm future people, even if they do not exist yet. A real-world example might be nuclear waste that kills an unsuspecting person who digs it up thousands of years from now.[4]

Like others had before him, Parfit highlighted that there could be an enormous asymmetry between human history to date, and the long, vast trajectory that could lie ahead. 'Civilisation began only a few thousand years ago. If we do not destroy mankind, these few thousand years may be only a tiny fraction of the whole of civilised human history,' he wrote. 'If we compare this possible history to a day, what has occurred so far is only a fraction of a second.'[5]

For Parfit, it therefore followed that those who are alive today, in this century, hold a great responsibility. He proposed that we may be living at a time when humanity's power to shape the lives of future people has never been greater. 'We live during the hinge of history,' he wrote. 'Given the scientific and technological discoveries of the last two centuries, the world has never changed as fast. We shall soon have even greater powers to transform, not only our surroundings, but ourselves and our successors.'[6]

What we end up leaving for these descendants could be a harmful legacy – the glass in the forest – but it needn't be: it could also be a positive legacy that allows future people to flourish and become the best that humanity can be. 'Life can be wonderful as well as terrible, and we shall increasingly have the power to make life good. Since human history may be only just beginning, we can expect that future humans, or supra-humans, may achieve some great goods that we cannot now even imagine . . . Some of our successors might live lives and create worlds that, though failing to justify past suffering, would have given us all, including those who suffered most, reasons to be glad that the Universe exists.'[7]

In a nutshell, we do not just have moral obligations towards people that are geographically far away, but also those who are *temporally* distant too. The circle of empathy and responsibility should extend across time as well as space.

Along the way, Parfit would discover that doing this is nowhere near as easy or straightforward as it first appears; nor is it as simple as the tale of the child in the forest might suggest. The bottomless pit he had started digging also held some undesirable conclusions and unanticipated consequences (some of which we'll return to in a few pages).

But by looking at humanity through the long-term lens that he did – reframing the inhabitants of the deep future as far more important than had been considered before – he would lay the ground for others to pick up the baton. This new generation of thinkers would take his theoretical arguments, build upon them and ultimately turn them into a movement.

The altruistic roots of longtermism

If the longtermists had all got together at a party in the late 2000s, you might have struggled to see how their interests overlapped.

In one corner of the party, you'd find the people who studied artificial intelligence, having a lively debate with the transhumanists about the threats and benefits of superintelligent machines. In another, there'd be the economists, rationalists and mathematically-minded, trying to probabilistically work out the likelihood of their hangovers. At the buffet table, you might encounter the decision-theorists, immersed in the consequences of their choices. Helping out at the bar would be the altruists, and on the roof would be the astronomically-minded, wondering if the stars above held other civilisations. Finally, down on the dancefloor, you'd see the existential risk crowd, letting their hair down after spending all day thinking about humanity's extinction.

But while they all ostensibly did different things in their day jobs, a few traits united them: for one, most were focused on analysing the future, and plenty were quantitative thinkers. Philosophically, many also believed in the importance of maximising happiness for humanity as whole, and the idea that one's actions should be decided by what leads to the most good. In other words, they were more likely to be 'consequentialist' in their ethical outlook, guided by the consequences of their actions rather than specific rules about right and wrong. And finally, a significant proportion were inclined to put these theoretical ideas into practice, as a guide to how they live their lives, donate their incomes and prioritise their careers – even if it led them to counter-intuitive conclusions or tricky personal decisions.

One of those hypothetical party attendees was the philosopher Toby Ord – and he was about to make one of those tricky decisions. Ord would not have described himself as a longtermist back then, because the word didn't exist yet. But along with several others, he was planting the seeds for a new, far more long-minded way of thinking, inspired by the ideas of Parfit.

In 2010, Ord was in his early thirties. Raised in Australia, he had moved to the University of Oxford to pursue philosophy. Studying under Parfit, a career in conventional academia lay ahead.

But one day, Ord made a decision that would prove to be enormously influential – he publicly committed to donating more than £1 million of his money to fighting poverty. Curious and mildly incredulous journalists from the BBC, *Wall Street Journal* and *Daily Telegraph* asked him how he'd do such a thing on a university salary. After all, he was no Bill Gates. He explained that he had calculated that he could comfortably live off a personal allowance of around £330 ($450) per month after paying tax, rent and putting aside some savings, so he planned to give away the rest.[8] He'd continue donating at least 10 per cent of his income every year for his entire life (he actually settled on a higher figure). Writing their profiles, the journalists noted the signatures of this modest lifestyle: at the time, Ord lived in a sparsely furnished one-bedroom flat with his wife, a junior doctor, who had made a similar commitment. He dressed simply, went to dinner once a fortnight, and bought a coffee once a week.

Although his academic salary was unremarkable by Western standards, Ord explained that he had realised that he still sat among the world's most wealthy people, in the top 4–5 per cent. And crucially, that spending a pound of his money in the developing world would go far further than if that pound was spent in the West. You can save somebody's life in rural Africa by donating only a few thousand pounds: funding charities that provide anti-malarial bed nets, for example.

Ord was influenced by his supervisor Parfit, who had argued that the affluent have strong moral obligations to the poor. But also the philosopher Peter Singer, who had proposed a famous thought experiment called 'the drowning child in the shallow pond'. It asks you to consider whether you would be willing to wade into muddy, dirty water to save the life of a desperately struggling toddler if it meant destroying your new pair of shoes and damaging your suit. Few people would say no. So Singer asked: how is that different to spending some of your disposable income to save a child from a deadly disease in a far-away land?

Ord has given away a significant proportion of his money ever since, as well as encouraging thousands of others to donate at least 10 per cent of their salaries, via the Giving What We Can Pledge. Singer pledged the same, as did Parfit.

What Ord did – along with Singer and others at the time – was show that it's possible to take abstract principles and thought experiments, and apply them practically in the real world to try to do good. It was an approach that, as we'll see, would be central to longtermism too.

First, though, Ord's experiences would lead him to establish what would become an influential social movement, along with William MacAskill, another young Oxford philosopher. The story goes that the pair met for the first time in a graveyard within the gardens of St Edmund

Hall, one of the Oxford colleges, and went on to talk for hours. Out of this conversation within a territory of death sprung an ambitious idea for saving life. They called it 'effective altruism'.

Effective altruism aims to apply philosophical principles and reasoned evidence to help people maximise the good they can do with their money and time. It has successfully redirected hundreds of millions of dollars to programmes that, for example, provide anti-malaria mosquito nets or the deworming of children, and while it is an approach that has its critics, it has helped to save many, many people's lives among the world's poorest populations.[9]

So, what would draw Ord, MacAskill and others to longtermism? What does charity in rural Africa have to do with the deep future? It began with a simple question: if we have the power to save the life of a person in the present, shouldn't that obligation extend to people in the future too?

One of the turning points was a 2013 doctoral dissertation by the philosopher Nick Beckstead, titled: 'On the Overwhelming Importance of Shaping the Far Future'.[10] Beckstead had picked up the baton left by Parfit, and combined those ideas with arguments about how to maximise the good you can do. Over nearly 200 pages, he made the case that what matters most is to aim to do what is best for future generations over the coming millions, billions and trillions of years. This was further than Parfit himself went. Beckstead was not just arguing that the long term mattered; he was suggesting that it could be the most important thing. Not everyone would align with this stronger version of the argument, but the net effect was to place the ideas of the long-term future more firmly on many people's radars.

It also came at a time when many researchers had begun to think about the long term through another lens: future existential catastrophe. A number of influential papers published around this period raised concerns that we may be hurtling towards the extinction of humanity through a human-made threat, such as artificial intelligence, a bioengineered pandemic or extreme climate change. Before then, the study of the end of the world had tended to be more of a theological issue, but when researchers like Oxford University's Nick Bostrom painted the chilling trajectories that could lay ahead, it established a whole new field of study. Amid this context, down the corridor from Bostrom at Oxford's Future of Humanity Institute, Toby Ord had begun to build on Parfit's ideas about existential risk and the hinge of history. This would turn out to be very relevant to longtermist thinking.

Parfit had once proposed a simple thought experiment, where he asked you to consider the following three outcomes:

1. Peace.
2. A nuclear war that kills 99 per cent of the world's existing population.
3. A nuclear war that kills 100 per cent.

What is the difference between each scenario? Obviously, peace is far better than 99 per cent of the population dying. But what about the difference between the latter two? Both would be awful, but most people when asked tend to see them as similarly severe. If 99 per cent of humanity dies, that is pretty much as bad as 100 per cent, right? If you agree, you wouldn't be alone: researchers have found a majority of people take this view when asked.[11]

For Parfit and Ord, this conclusion is wrong and massively underplays just how much more terrible scenario three would be. If humanity goes extinct, you don't just wipe out all human life today – you lose the potential for *all human life there could ever be*. If we were foolish enough to cause our own extinction, it would therefore be an almost unimaginable tragedy. All the flourishing, well-being, love, happiness, connection and achievement that lies ahead for humanity would be gone. 'I believe that if we destroy mankind, as we now can, this outcome will be much worse than most people think,' wrote Parfit.[12]

Ord would build on these ideas by analysing all the different ways such an extinction could happen, trying to probabilistically estimate their likelihood. He looked at natural causes like asteroids and supervolcanoes, but also human-made disasters like nuclear winter, or an AI catastrophe. He concluded that we may now be living through one of the most precarious periods ever. We are the first generation with the technology to destroy ourselves, he realised, but we have not yet developed the wisdom to ensure we don't. Ord would call this period 'the precipice'.[13] It began on 16 July 1945, the date of the first atomic weapon test – and since then, our societies have only added to the list of human-made threats.

Through this work, Ord continued as Parfit had begun, motivated by the belief that we owe it to future people not to prevent their existence. 'We stand poised on the brink of a future that could be astonishingly vast, and astonishingly valuable,' he argued, and therefore we have a moral duty to ensure that we don't take this future away from our descendants.

Longtermism takes off

Gradually, more and more people were drawn to thinking more deeply about our ethical obligations to the future of humanity. It was a lens on the world that would unite all the disparate groups at the party. For the effective altruists, it wasn't too much of a leap to wonder whether their money could help to save lives in the future as well as the present. For the probabilistic predictors and decision-theorists, the dimension of deeper time offered them new ways to consider future acts and consequences. It gave the artificial intelligence crowd a bigger cause: avoiding catastrophe so that humanity's extremely long future could be preserved. And for the more sci-fi oriented, it provided a more rigorous intellectual framework to talk about the potential for galactic settlement, or how tomorrow's people might change through transhumanism – a movement focused on evaluating how human beings could be enhanced through technology.

A name was therefore needed to assemble around. Beckstead had written about the 'far future', but Ord suggested that the word 'longtermism' was better. 'The long-term future includes all the times from now onwards, rather than just gesturing at a point a long way off,' Ord told me, 'and it is this broadening which explains why it is so important: there is only one year a million years hence, but there are a million years leading up to it.' (His point aligns with the psychological effects we explored in Part II, where the word 'far' becomes unhelpfully distancing.)

These longtermists would go on to question many assumptions about the future, and the priorities we hold in our societies. One of their early targets was to call attention to a particularly time-blinkered practice called 'discounting', which is embedded in governance across the world. The reason the longtermists so dislike it is because, when it is applied without care or thought for the long-run consequences, it promises to shape people's lives in the future profoundly.

What exactly is discounting, and why does it matter? It's not a particularly intuitive concept, but it is worth explaining, not least because it underpins so much of the short-sighted economic and political decision-making we explored earlier in the book. So let's briefly pause our story about the rise of longtermism to imagine a newly elected politician – let's call her Clarissa – who has a dilemma.

A discounted future

One day, the British prime minister Clarissa and her Cabinet were weighing up whether to spend a few billion pounds on a giant asteroid detector that could one day save many lives. Her scientists have assured her that, one day, it could be needed urgently, and her great-grandchildren would be grateful. But the upfront cost would be painful, so she was conflicted. Her constituents were crying out for investment, the media was hostile and she had been elected by promising tax cuts.

One of Clarissa's civil servants had a suggestion: he whipped out the UK Treasury's Green Book – its guidance on how to appraise policies, programmes and projects. Inside, there was a number called the 'discount rate'.

A junior Cabinet minister piped up. 'I don't see what discounts have to do with this. A shop sale? "Now with 20 per cent off"?'

'No, no, think of it as the verb, to devalue or disregard,' the civil servant replied. 'As in, "I decided to discount the minister's advice." Look, it's standard practice when we're making decisions about the future. Essentially it means that we needn't value economic benefits down the line as much as value them today. Don't worry, governments all over the world do it!'

Discounting, he explained, is an idea that can be traced back to the 1800s, when the Scottish-Canadian economist John Rae observed that both people and nations tend to weight short-term rewards in the present more heavily than gains in the intangible future. Rae saw this bias towards the present through an economic lens, and his work laid the foundation for the idea that it's possible to *quantify* it.

In the 1930s, Rae's suggestion that people have a 'time preference' was fashioned into an economic model by Paul Samuelson, one of the most influential economists of the twentieth century. In a five-page article, Samuelson aimed to capture how a society's perception of future value diminishes over time, calling it the discount rate. Since then, discounting has evolved significantly, but the underlying principles have transformed our societies and their governance, with far-reaching effects to this day.

Economists argue that time preference exists – it's human nature – and that if we didn't place at least some extra weight on the present, then we would be obliged to spend almost *all* our resources on the future. It'd be like asking you to live a life of extreme frugality, putting every spare penny or cent you earn into a pension fund – it's just too demanding. It's therefore necessary to have some discounting, in order to maintain and improve the present-day standard of living. Economists also point out that if economic growth continues over time, it will mean tomorrow's societies will be richer and thus better able to bear costs.

So, Clarissa and the civil servant took out their metaphorical calculator to crunch the numbers. This showed that investing in the asteroid detector would have plenty of economic benefits – the nation's businesses will survive, for starters – but also that these upsides won't kick in for many,

many years, possibly even centuries. Since they were so far away in time, the discount rate devalued these benefits to be much lower than if they happened tomorrow. So, it was better to spend the money on something with more immediate impact. Seeing the sums, Clarissa chose to leave it to her successor to build.

While Clarissa's asteroid detector example is an exaggeration – the decision to save the world from a giant rock from space probably wouldn't be made this way – many other major policies *are* shaped by the practice of discounting, from whether to build transport infrastructure to investment in healthcare. The UK Treasury's Green Book is real, with a standard discount rate of 3.5 per cent each year, and 1.5 per cent for policies that impact health or life outcomes. It does decline over time – but is never zero.[14]

Discounting can make sense when applied narrowly to relatively short-term economic decisions, ostensibly helping policymakers decide whether, say, a public transport megaproject is worth the initial outlay. But the practice can also support errors of judgement and short-termist cost-cutting like building weaker bridges that go on to collapse in a few decades' time. And it becomes particularly problematic when applied to policy choices that devalue people's future welfare over longer timescales.

This was clear even to the economists who developed the first discounting methodologies, a century ago. In 1928, British philosopher Frank Ramsey introduced a mathematical framework for discounting, but in the same paper cautioned that devaluing people's 'later enjoyments' within these calculations would be 'ethically indefensible and arises merely from the weakness of the imagination'.[15]

These ethical implications of discount rates have been at the root of vigorous debates about climate change – and how urgently to make investments in prevention and mitigation as the effects rapidly worsen. Most serious economists now accept that there's a need to bear some costs to avoid climate catastrophe down the line. But how much spending is acceptable, and how quickly? When economists are debating this question, they are essentially arguing over how big a discount rate to apply.[16]

But if the environmentalists disapprove of discounting, the long-termists are even less of a fan. They point out that if you apply a discount rate to people's welfare over *really* long timescales – say 1,000 or 10,000 years – it eventually devalues the lives of future people to *almost zero*.

This leads to conclusions that are not just ethically indefensible but seemingly absurd, suggesting that the welfare of one person today could be worth the same as millions of lives in the future. This was Parfit's point about the child and the glass shard in the forest. The two children may seem equivalent, but discount rates would suggest that the present-day child's welfare was worth more. He also pointed out that if Ancient Egyptians had discount rates – and that they had the power to make decisions that affect our welfare in the present – then your life's value would be worth almost nothing to them. 'Imagine finding out that you, having just reached your 21st birthday, must soon die of cancer because one evening Cleopatra wanted an extra helping of dessert,' he wrote.[17]

There are ways to solve some of these problems – applying a declining discount rate over time is one. The UK Treasury's Green Book does this, as well as recommending different approaches for policies that have health and life outcomes, or environmental consequences. But for many, this doesn't go far enough.

In 2021, Ord and the Centre for Long-Term Resilience co-authored a report in which they called for revisions to the discounting used in the Green Book.[18] The present figure, which begins at 3.5 per cent per year, is too high, they argued, and should decline more quickly than is recommended. They also point out that the Green Book fails to acknowledge the costs and economic harms of significant disasters that could harm future generations, and needs more detail on how to account for second-order effects of policies.

But calling attention to the problems with discounting was really just the beginning for the longtermists. As the ideas began to bed in, and other philosophers started weighing in, longtermism moved into new territory – and it raised some profound implications that were not without some controversy.

The scale of the unborn

When Derek Parfit wrote that humanity was only at the beginning of a potentially very long future, he did so in broad terms. But if you were to put some numbers to it, just how many future people are we talking about?

A few years ago, I performed some rough calculations for the BBC to get a handle on the scale of future generations. If you assume, for the sake of illustration, that the birth rate were to stay at this century's average and the global population were to remain stable for the next 50,000 years, there stand to be 6.75 trillion more people born – which is more than sixty times the number of humans *that have ever lived*.[19] If each of those unborn people was represented as a letter in this book, you would need a print run of almost 17 million copies to flick through them all.

Here is what that looks like, illustrated to scale:

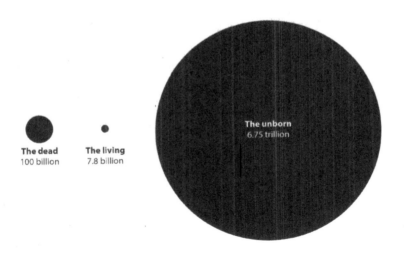

My calculation was subsequently picked up and more widely shared, in particular by the author Roman Krznaric in his book *The Good Ancestor*. He used the diagram to propose a set of scales carrying all present people on the one side, and the unborn trillions on the other. 'How could we possibly ignore their well-being and think that our own is of such greater value?' he wrote.

But enormous as 6.75 trillion is, I now realise that I probably under-estimated it. The scale of future generations could actually be far, far bigger.

For starters, our species has the potential to last far longer than 50,000 years. In fact, the longtermists believe that there's every reason that our descendants have the potential to last for millions, if not *billions*, of years. If those great-grandchildren in the deep future were to stay on Earth until the Sun dies, then the total number of unborn people could exceed *100 trillion*, according to one ballpark estimate by Oxford University's Toby Newberry, who used a different set of assumptions to my own.[20] That's almost 1,000 times more than all the people who have ever lived.

Hang on though. Billions of years? Plenty of people might expect a far more imminent demise for humanity than that – we'll be lucky to get through the next century, they might say. The longtermists do recognise that we are living through a particularly precarious period – it's why Ord calls our time 'the precipice' – but they also believe that if we can work out how to reduce the anthropogenic risks we face over the coming decades, then the future beyond could be extremely long.

Going extinct now, so early in our evolution, would actually be quite unusual for a mammalian species, which typically endure for an average of one million years. *Homo sapiens* is only a third of the way there. The sceptic might respond that other animals can't annihilate themselves with their technology. True, but we mammals also have intelligence, language and foresight to invent, communicate and anticipate our future, and our species is also far, far more widespread across the planet's surface than any other large vertebrate, occupying every corner of the Earth.

I realise that you may still not be convinced we have a billion-year future. But as Newberry puts it: 'The claim that our future will be anything shorter than a million years, or contain any fewer than 100 trillion lives, in expectation, would require both extraordinary scepticism about these grander futures, and extraordinarily high confidence that humanity will cause its own destruction in relatively short order.'

By some longtermist estimates, the figure of 100 trillion is not even the upper limit. There are calculations that go further – quite a bit further actually, so buckle up.

Firstly, if our descendants pursue a so-called 'astronomical trajectory', expanding out into the cosmos, then the numbers get far bigger. If they settle the Solar System, the number of lives could be 10,000,000,000,000, 000,000,000,000,000 (10 octillion). If they spread out across the Milky Way? Add nine more zeroes. And add another nine if they settle the affectable Universe beyond.

Some longtermists – who you might label the techno-utopian faction – even add future 'digital minds' into their reasoning. These would be 'people' whose consciousness exists only in silicon. The most well-known number comes from Oxford's Nick Bostrom, who once calculated that: 'at least 10^{58} human lives could be created in emulation even with quite conservative assumptions'.[21]

(It should be said that these sci-fi-tinged visions are not universally embraced. The core arguments of longtermism do not rely on inter-galactic diasporas or digital people, and some longtermists think that talking about such speculative futures distracts from risks we face much closer to the present.[22])

Billions. Trillions. Octillions. Such numbers are so large it's difficult to imagine them, so let's try to understand them in physical terms. Imagine each page of this book was a person.[23] To represent the current population of Earth, it would have to be roughly 780km (485 miles) thick, which is the drive from London to Marseille. The book of 100 trillion people would therefore have to be 10 million kilometres (6,213,712 miles) thick, which is twenty-six times the distance to the Moon, and around 250 times the circumference of the Earth at the equator.

To bring it back down to more tangible terms, now imagine one page represented *all* people alive on Earth today. To reach 100 trillion, you would need to leaf through almost 13,000 pages. That's about 10–11 Christian Bibles, 24 *Great Expectations* or 300 *Communist Manifestos*.[24]

Or to deploy an appropriately temporal metaphor, you could also imagine each person was a second ticking by on the clock. To count to 100 trillion, you would have to wait 3.2 million years, which happens to be the same length of time between us and Lucy, the famous Australopithecus bipedal hominin.

When you get to octillions, though, it gets even crazier. You can't meaningfully represent it with book pages, and one octillion seconds wouldn't fit between now and the Big Bang, so another metaphor is required. If you poke around the websites of amateur mathematicians, you can find various analogies. The Earth apparently has a mass of about 5.98 octillion grams, for example.[25] But my favourite way to parse the size of an octillion involves a raining vegetable:

First, imagine if a blizzard of peas fell from the sky, coating the entire land surface of Earth, so that all the continents were buried to 1.2 metres (4 feet). This blanket would contain a sextillion peas (21 zeroes).

That's not enough.

So, next imagine the oceans froze over, and the peas fell on top of that surface too. That would get you to a septillion (24 zeroes).

Still short.

To get to an octillion peas? You would need to rain them over 250,000 more planets the size of Earth.[26]

Strong longtermism

It is these mind-bendingly massive numbers of potential future people that inspired the philosophers Hilary Greaves and William MacAskill to take longtermism to a whole new level. They call it 'strong longtermism'.

Like many other longtermists, the pair started out in different fields. Greaves began her career in the philosophy of physics, centring her work on the interpretation of quantum mechanics and the 'many worlds' theory that multiple universes exist. MacAskill began his own helping Ord to establish the effective altruism movement – he also made a lifetime pledge to give away at least 10 per cent of his income – and made his name as something of a philosophy wunderkind, becoming an associate professor at Oxford at the age of only twenty-eight. He has since become one of the most visible and influential proponents of longtermism, and at the time of writing, was poised to publish a book making a case for the movement called *What We Owe the Future*.[27]

When MacAskill and Greaves combined the ideas of Parfit, Ord and others with estimates about the vast size of future populations, they would show just how far longtermism could go in terms of shaping our present-day priorities. Strong longtermism argues that if humanity's future could be as vast as it seems it might be, then our most important actions today are those we take to protect it and improve the lives of future people. If we can find reliable ways to work towards ensuring that this huge population exists and flourishes, then this should be a very high priority.

Therefore strong longtermism goes much further than asking politicians to adjust economic habits like discounting, or broadly seeking to ensure the future goes well. It makes the case that, if we want to do the most good we can, a much larger proportion of time and money should be spent on causes that have far-future benefits.

So what might that look like in practice? Mainly, Greaves and MacAskill choose to look at it through the big-picture lens of philanthropy and governmental funding, rather than concerning themselves with personal day-to-day decisions. For example, consider what strong longtermism says about the possibility of a human-made viral pandemic. To date, governments have spent relatively little on measures to reduce this risk, despite the fact that several laboratories in the US, Russia, Japan and elsewhere have fiddled with viruses to study or weaponise them. As Ord once pointed out, the annual budget of the UN Biological Weapons Convention is less than the average McDonald's restaurant. It's chilling to consider what could have happened if the SARS-CoV-1 virus had been deadlier than it was, a deliberately engineered version with a far higher fatality rate.

Spending some money on efforts to prevent such a killer virus might already seem like a no-brainer. But Greaves and MacAskill say the case becomes far stronger if you also factor in the impact on future people. Some estimates have suggested that we could reduce the likelihood of this path to extinction by around 1 per cent by spending just $250 billion on the problem.[28] That may not seem like a huge percentage, but Greaves and MacAskill say, on the contrary, making investments like this one offers the biggest bang for buck that there is. If you want to save lives, they say, you couldn't spend it better. Why? Plug in the numbers of potential Earth-based life, and it suggests that each $100 of that $250 billion could increase the number of future beings by 200 million. According to this logic, there is no more effective way to cause more human beings to flourish.

Greaves and MacAskill are making the case that we *should* do this, morally, but they also carefully present the pragmatic argument too. Morals aside, if we merely *want* to have the most impact, then the future is where the majority of the value is. So if your goal as a politician or philanthropist is to invest money with the most effective results, they say, put it into something that improves the lives of all those trillions of people in the deep future.

Most academic philosophy papers that deal with such questions are simply published in a journal and occasionally get debated in a university lecture hall. But arguments like Greaves and MacAskill's are already having tangible effects on how money in the wider world is allocated, inspiring a number of philanthropic organisations and donors to direct funds towards causes that could benefit future people. As of 2021, the foundation Open Philanthropy had already given one third of a billion dollars to longtermist cause areas (and they also, for the record, have supported my own work; OpenPhil funded a research assistant to help me with this book). This doesn't mean that these organisations have stopped funding efforts to help people in other ways, such as charities that provide anti-malarial mosquito nets in the developing world, but now they also back pandemic preparedness and work to prevent other potential global catastrophic risks, like rogue artificial intelligence.

What began, then, as a philosophical idea has rapidly become a movement, inspiring thousands of people to focus their time and efforts on longtermist research areas. Websites like 80,000 Hours advise people on how to choose a career path that will benefit the deep future, and it's becoming ever-more common for people to self-identify as a longtermist in their outlook and beliefs. At the time of writing, it was mainly a phenomenon within the world of academia, the tech industry and research organisations, but the pace of growth already has been dizzying to observe.

How to be longtermist

So, what should the rest of us make of all this? If one wishes to be more long-minded as an individual, does that then oblige us to follow the conclusions of longtermism?

Firstly, it's worth pointing out that there are already differences of opinion emerging within longtermism about what to prioritise, and just how urgently to act on its conclusions so far. It is, after all, still a nascent idea. For example, not all longtermists share ground with the techno-utopian faction who talk of redirecting humanity's path towards specific technological and astronomical futures. This more bullish approach within the movement has already attracted criticism for its supposed detachment from present-day suffering and risks like climate change.[29] For others, longtermism is simply about spending on prudent causes that reduce the chances of catastrophe: investments that could benefit ourselves, our children *and* people in the deep future, such as reducing the existential dangers of nuclear war, pandemics, AI, climate collapse, or asteroids. Longtermism in this spirit is closer to shelling out for a helmet to ride a bike. Yes, there's an upfront cost, and the likelihood of you dying in a bike crash is small – but you take the precaution.

But more generally, the extent to which you will buy into the ideas of longtermism probably depends on your own ethical framework. The arguments of strong longtermism are on much firmer ground if you are a consequentialist and believe that doing good is about maximising happiness for the maximum number of people. If you are not, however, the moral case for strong longtermism may leave you cold.

To test their argument, Greaves and MacAskill do make a number of assumptions. For example, they largely disregard what is called the 'person-affecting' view. If you are person-affecting, you believe that the well-being of actually existing people matters more than hypothetical people who don't exist yet. To you, adding more people to the world is neither good nor bad. As one philosopher once put it, shouldn't our moral goal be to *make people happy*, rather than *making happy people*?[30] Parfit too wrestled with this dilemma.

Then there are those who worry that longtermism could be taken too far. Even within the specialist circles where longtermism is discussed, some have raised concerns that these ideas could evolve into a cause that could consume all others in terms of attention and money.[31] If our societies were to become 100 per cent strong longtermist, wouldn't it suggest that we should therefore focus almost *all* our available resources on producing a flourishing future? As philosopher Jim Holt writes, an extreme longtermist viewpoint might lead someone to conclude that 'at the limit, we should direct 100 per cent of our time and energy toward protecting humanity's long-term future against even the remotest existential threats – then wrap ourselves in bubble wrap, just to be extra safe'.[32]

And what about present-day problems such as poverty, refugee crises, or even climate change? While strong longtermism does not argue that such issues should be ignored, there are critics who nonetheless see it as a form of mathematical blackmail, where the huge scale of future people could be used to justify demanding and perhaps even harmful sacrifices in the present, in the wrong political hands. This 'bad actor' fear does of course apply to many politically oriented '-isms', from libertarianism to socialism, and so it does not mean the entire system of ideas is dangerous, but the view that longtermism could be used as a justification for harm or neglect in the present is, for some at least, a strongly held one. Even

Peter Singer – a philosophical icon to many longtermists – has raised this as a potential danger.[33]

The apparently demanding nature of strong longtermism is not something its proponents are blind to. The consequentialist philosophy underpinning these ideas has faced such objections before, characterised by thought experiments like the so-called 'happiness pump'. This idea was recently captured in human form by the character Doug Forcett in the TV show *The Good Place*. Forcett believes that it is his life's duty to increase the happiness in the world, so spends his whole life doing everything he can to please others, even though it makes him thoroughly miserable. He lives a monastic life of extreme frugality, gives away his time and possessions to anyone who asks and even allows a teenager to relentlessly bully him. Imagine if Forcett was a strong longtermist: he would feel obliged to detrimentally sacrifice himself for the sake of future people's happiness as well.

Another challenge, particularly for the most interventionist longtermists, is that most things about the future are unknowable. Preventing extinction is one thing, but those who would seek to steer humanity towards a singular trajectory may well be acting in the dark. Over very long timescales, it's difficult to say if what seems like a positive action today could actually be a waste of time and money, or worse, negative for future people.[34] Consider that if longtermism had emerged in Ancient Greece, philosophers might have assumed that the best thing for future generations would be to ensure the balance of four humours: blood, yellow bile, black bile and phlegm. If it had emerged in Europe in the Middle Ages, many might have decided that the number one cause to support was a religious crusade. The point is that over long-term time, our scientific knowledge, technologies and even our moral values will change – and it's difficult to predict how.

With these thoughts, we're getting lost again in Parfit's 'bottomless pit of doom': contradictory conclusions and no easy answers. Longtermism is a relatively new philosophy in the history of ideas, and its strongest form is already proving divisive. It may take more years to refine – and perhaps even another generation of thinkers to take the baton down the line.

I don't know if Parfit would have described himself as a strong long-termist, because he died in 2017. But MacAskill chose not to defend the strong version in his book for a general audience, suggesting that 'it is not a view that we should be highly confident in'.[35] And Ord's preferred approach to longtermism is more akin to environmentalism: 'It doesn't mean that the only thing that matters is the environment, just that it is a core part of what you care about and informs a lot of your thinking,' he has said.[36] 'Longtermism . . . takes seriously the fact that our own genera-tion is but one page in a much longer story, and that our most important role may be how we shape – or fail to shape – that story. Working to safeguard humanity's potential is one avenue for such a lasting impact and there may be others too.'

If this is what longtermism means, I can see it growing to become a more widely held view across society, joining up with the other forms of long-term thinking. I'm less confident that the same could be said of the more extreme mathematical iterations of the argument, especially if its messengers also came talking about digital minds, technological futures or intergalactic colonisation. Personally, I can get behind the general prin-ciples of longtermism, but not the strongest or techno-utopian forms. As you may have gathered by reading this book, I have been influenced by the ideas of Parfit, Ord and others. I believe that the future could be far longer than we currently imagine, and that we could do much more to ensure that tomorrow's generations have good lives. I also think that it makes sense to invest in prudent precautions that reduce the likelihood of a global catastrophe or human extinction. However, long-mindedness, to me, is broader than a calculation about the hypothetical unborn, and should be grounded in the relationships we hold today and with our predecessors.

There's also one more reason why I struggle to fully embrace strong longtermism: it is a timeview that talks a lot about humanity's future potential, but so far, less about nature's. Many effective altruists care deeply about improving animal welfare, particularly in factory farms, and have dedicated much of their personal income and careers to tackling this issue. But so far at least, it's difficult to see how to integrate concerns for nature and the non-human world into calculations that involve the well-being of future people.

This century, we face a pressing need to develop a more harmonious, long-minded relationship with natural systems, our fellow animals and the Earth. We've already touched on how the *environmental timeview* of non-industrial cultures can offer this perspective, so let's now round that out with insights from the *scientific timeview*. Those who study the non-human world have revealed a profound picture of our place within nature, as well the methodologies and knowledge we need to improve our long-term relationship with living and physical systems. What, then, might we learn from the world's most long-minded natural scientists?

10

Temporal Windows:
Science, Nature and the Anthropocene

'The strength of the thread does not reside in the fact that some one fibre runs through its whole length, but in the overlapping of many fibres.'

LUDWIG WITTGENSTEIN[1]

'The love of nature is different from the love of science, though the two may go together.'

JOHN BURROUGHS[2]

One of the most astonishing scientific facts I have ever heard is this: you and I are continuous with non-life.

How so? The evolutionary biologist Stephen C. Stearns once explained why in a class he taught at Yale University:[3]

Think of your mother. Now think of her mother. Now think of your mother's mother. Keep going back in time . . . speed it up . . . we're at 10 million years, now 100 million, a billion years. Every step of the way there has been a parent.

At 3.9 billion years ago, something extremely interesting happens.

You pass through the origin of life, and there's no parent any more. At that point you are connected to abiotic matter.

Now this means that not only does the tree of life connect you to all the living things on the planet, but the origin of life connects you to all matter in the Universe. That's a deep thought. Every element in your body that is heavier than iron, and you need a number of them, was synthesised in a supernova.

So, when people say we are all made of stardust, it's more than a poetic cliché: it's true. Or more specifically, the atoms that make up 97 per cent of the human body are also distributed across the galaxy.[4] It just happens to be that these elemental building blocks are arranged in a way that makes you, *you* – and indeed, every other living thing on the planet.

We still don't know if the origin of life on Earth was the only place in the Universe that it happened. But the fact that it did, after billions of years of sterility and barren chaos, made it an event of such long-term importance that it's difficult to adequately describe in language. One of my favourite words for it is a *eucatastrophe*: a coinage by J. R. R. Tolkien in 1947 to describe a 'joyous turn' within a fairy story; a 'sudden and miraculous grace: never to be counted on to recur'.[5]

This long view of life's efflorescence and kinship is just one example of how science has helped to reframe humanity's position within time and nature. But it was a perspective that was not easily won: it took hundreds of years to emerge, and it's almost certainly not the final word either. There's still so much we don't know about our long-term role in the biosphere and nature's deeper trajectory. The remarkable thing about science is that it is never finished. As the astronomer Carl Sagan wrote in 1979: 'Science is a way of thinking much more than it is a body of knowledge. Its goal is to find out how the world works, to seek what regularities there may be, to penetrate the connections of things.'[6]

What might this 'way of thinking' teach us in our quest for a richer form of long-mindedness? Science covers many realms of inquiry – but it offers a particularly profound and illuminating timeview when it comes to understanding our role in the natural world, so that will be our focus in the coming pages. After all, nature is too complex for any single person to perceive all at once, and when you include its deep past and long future as well, the scope becomes vast. What are the intellectual approaches and methodologies that scientists use to overcome such challenges? And in turn, what might their discoveries tell us about the human relationship with nature over the long term?

Let's begin with a story about two scientists who, one day, noticed a puzzling signal from their telescope. At first, they thought it might be explained by pigeon poo caked in the dish – but it turned out to be something far older and grander.

Temporal windows

One day, the astronomers Arno Penzias and Robert Wilson were troubled by a noise. This was back in the 1960s, and the pair were operating a small radio antenna in the town of Holmdel in New Jersey. Pointing the antenna at the sky, they had been trying to detect radio waves from distant cosmic objects, but there was a background hiss in their readouts that they could not get rid of.[7]

Could it be the sounds of nearby New York City? They ruled that out. Natural noise from the ground or atmosphere? Probably not. Might a fault on the edge of their antenna be causing the problem? They solved that with sticky tape.

What about the pigeons roosting in the telescope dish? Surely it wasn't the birds, but they had to investigate. So, methodically and messily, they climbed into the antenna and scrubbed off all the pigeon faeces. They also caught the pigeons, and dispatched them humanely to a location far away. Even then, after all that cleaning, the noise remained.

In December 1964, Penzias was travelling on a plane from a conference, still vexed by the hiss, so he mentioned it to a fellow astronomer. A little while later, the astronomer got back to him with some intriguing news:

two physicists at Princeton University, he said, had recently published a prediction about the legacy of the Big Bang. They had been working only thirty miles away, but this was news to Penzias.

At the time, the Big Bang theory lacked the wealth of observational evidence it has today. But the Princeton physicists had suggested it might have left behind some background radiation – which might even be detectable with a radio telescope. In other words, the Big Bang would have left a small hiss.

Penzias and Wilson realised that they had accidentally discovered something truly ancient. It wasn't pigeon droppings. It wasn't interference from New York City. It was actually one of the oldest things that science has ever detected: the 'cosmic microwave background', also known as the Big Bang's afterglow. When visualised by present-day telescopes as a map of the sky, it looks like a Jackson Pollock painting, with splodges of blue, yellow and red to represent different regions, demonstrating that the Universe is far from smoothly arranged.

This electromagnetic radiation formed more than thirteen billion years ago and harks back to an era when the Universe was made only of hot plasma and photons. It permeates all of space – if you detuned an analogue TV, it would make up a portion of the static. And hidden within it are fluctuations that provide clues to how stars and galaxies formed – not to mention deeper mysteries, such as a huge 'cold' patch that some cosmologists believe could be evidence of a parallel universe, or circular spots that may have been left over from before the Big Bang.[8] Ultimately, what made Penzias and Wilson's discovery so important – winning them the Nobel Prize in 1978 – is that it gave scientists a form of long-term time travel, allowing them to understand the Universe as it existed in its infancy.

The cosmic microwave background is an example of what I call a *temporal window*, through which natural scientists can peer into deep timescales. Over the centuries, science has discovered many more of these windows: ice cores, fossilised bones, tree rings, stalactites, mineral inclusions, ancient pollen or mitochondrial DNA – to name a few. And you don't necessarily need specialised instruments to appreciate them.

Look down, and you might note the ancient sand of a beach between your toes, or the million-year-old rocks exposed at a building site in the city. Look up, and you can see starlight emitted long before *Homo sapiens* evolved. The Andromeda galaxy, for example, can be seen with the naked eye in some months of the year, but the light striking your retina took 2.5 million years to get here. The signatures of deep time are all around us, if we care to seek them out.

For scientists, temporal windows allow for the reconstruction of past worlds in a much richer palette. But their discovery can also lead to a richer understanding of the long-term *future* too. How so? We can see why by looking at another window – a curious magnetic pattern on the floor of the Atlantic Ocean – and what it told scientists about the deep-time trajectory of the continents.

In the 1960s, geologists noticed what looks like a huge magnetic barcode embedded in the rocks under the Atlantic seabed. In one stripe, the minerals in the rocks are magnetically oriented north, while in the next, they are oriented south. The bands are hundreds of miles wide, and thousands of miles long. Fly from London to New York and you'll pass over them, alternating back and forth.

The pattern offers access to times in the past when the Earth's magnetic field flipped. Compasses haven't always pointed north – every few 100,000 years, on average, the planet's polarity reverses – and the swings between these orientations are preserved in the ocean crust. But the reason the discovery of these temporal windows was so world-changing was because of what it confirmed about the planet's long-term trajectory.

The only way the pattern could have formed is if the seafloor itself is spreading. Halfway between the Americas and Europe/Africa, there's a ridge of subsea mountains where new ocean crust forms; a fissure between two tectonic plates called a 'constructive plate boundary'. It passes through Iceland, before snaking down to the Southern Ocean. Along it, upwelling magma fills a gap between two tectonic plates, feeding a conveyor belt of new crust on either side. The cross-section below shows, roughly, how the barcode pattern forms over millions of years:

US   ⇦   ⇨   Europe

When geologists discovered seafloor spreading, it confirmed suspicions that the Americas are gradually moving away from Europe and Africa. The Earth's crust, they realised, is made of separate plates that wrench apart and crash together, carrying the continents with them. This theory – plate tectonics – has since allowed scientists to model how the world map will change over the next few 100 million years too.[9]

There are a few different scenarios for how the plates will rearrange: if the spreading in the mid-Atlantic slows down and stops, the continents will reassemble to form Pangea Ultima, which will look like a massive atoll, surrounding what was once the Atlantic. On another path, the continents end up around the North Pole in a new supercontinent called Amasia; on a third, a new plate boundary tears up along Europe's western coasts and a landmass called Aurica forms, with Australia and America at its centre. But perhaps the more likely scenario is a new supercontinent called Novopangea. In this future, the Americas will collide with Australia and Antarctica, which in turn will crunch into China, India and the rest of Asia. It could look like this:

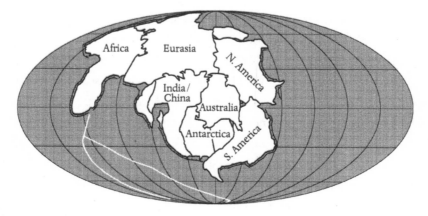

The projected arrangement of the supercontinent Novopangea.

So, in around 200–250 million years, our planet will look like an alien world. If our descendants are still around, they could travel between London, Los Angeles, Beijing and Sydney without seeing a coastline, and will have radically different neighbours to us. It's striking to consider how many wars have been fought over the possession of lands that, from the Earth's perspective, are just leaves floating on the surface of pond.

The reason we know this with such clarity is thanks to the construction of another tool that scientists use to unlock the long term: a *temporal model*. When geologists are seeking to understand the movements of the continents, they are doing so with modelling and simulation. And these models are often created and calibrated using data that comes from temporal windows (among other things). The techniques span many scientific disciplines: ancient ice cores help to build climate models, fossils enable evolutionary models, and so on.

What makes the model-based perspective so powerful is that it shows a range of long-term scenarios: possibilities for what happened in the distant past before humans came on the scene, and also what *could* happen across multiple futures. A common misconception is that models – particularly climate ones – are for making single predictions, like weather forecasts. But a climate model isn't designed to tell us if it will rain on 1 January 2100. It is for understanding ranges of possibilities and finding answers to 'what ifs'. If we do A, then B or C could happen; but if we do X, then Y or Z are also in the picture.

No model is a facsimile of the world, and some are better than others: they are shaped by the quality of the data you feed in, and the assumptions you make. But this is also the point: over time, they allow scientists to see if their descriptions of the world – past, present and future – are accurate. As the geologists behind the supercontinent simulations write: 'The objective of modelling the future is not just trying to guess what is going to occur but instead is a way of pushing the boundaries of our knowledge and trying to understand what the main processes operating at these long-time scales are.'[10]

Scientific models therefore bring us far closer to understanding how

complex systems operate than we could otherwise achieve. As the psychologist Tom Stafford once pointed out: 'Models are deduction machines . . . a
prosthesis for thinking, enhancing our perception beyond the horizon of
individual reason and intuition.'[11] But you don't necessarily need a super-
computer to embrace this way of thinking – at the risk of overstretching, I
would argue that a model-based timeview is essentially about being open to
the idea that there are multiple possibilities, past and present, and showing
a willingness to alter one's view as new information emerges. We all build
mental models about how the world works – without them, anticipation
would be impossible – but how often do we update or discard them?

So, science's temporal windows and models can reveal humbling truths
about our place in the natural world, allowing us glimpses into timescales that would otherwise stay hidden. This perspective shows that
the Universe will continue to expand and the continents will gradually
reassemble, no matter what we do.

However, this certainly does not mean that the natural and physical
world has its own separate path. On the contrary, the scientific timeview
has demonstrated that human activity has perturbed nature since our
earliest days on Earth. While there's no doubt that some societies and
peoples have been more responsible than others, the impacts of *Homo
sapiens* as a whole go back thousands of years. As we'll see next, this is a
long-term perspective that we need to understand if we wish to embrace
a richer and more holistic long view, not least because it puts the events
of the past century into stark relief.

The dawn of the Anthropocene

Our species has influenced evolution, ecosystems and the physical world
since the beginning. As far back as 50,000 years – as the human race started
to grow in size and expand across the globe – our ancestors were shaping
the composition of the biosphere. Fossils and other temporal windows show

that hunter-gatherers drove 178 megafauna species to extinction. The ecological niches left behind by some large mammals have never been filled.[12]

Windows such as ancient pollen have also allowed scientists to see that people began transforming natural landscapes up to 10,000 years ago. As agriculture spread, expanding populations began to gradually deforest swathes of land for crops and livestock. Many of the ecosystems that we take for granted as natural were, in fact, artificially created. For example, the bucolic chalk grassland of southern England was the result of forest clearing in Neolithic and Bronze Age times. Indeed, much of the continent of Europe was once wooded.[13] Meanwhile, climate reconstructions based on ice cores hint that the land clearance and emissions of early farming may have affected the atmosphere – nowhere near as much as today, but perhaps just enough to stabilise the climate and delay the next ice age.[14]

Once agriculture took off, selective breeding also began, creating new forms of life. By tracing mitochondrial DNA back through lineages, scientists can genetically time-travel to see the points that cultivated plant species such as wheat or rice, or domesticated animals such as dogs or livestock, were bred from wild ancestors. The domestic chicken, for example, came from a mixture of junglefowl that lived in India and China more than 5,000 years ago. Today, the chicken is the most common bird on Earth: at any given time, there are more than twenty billion of them – so fast have their numbers expanded that their bones will be clearly visible in the geological record.[15]

As soon as our ancestors began to travel more widely, they also began mixing up natural ecosystems. Across seas or mountain chains, humans often carried species with them – deliberately so, but also as unwitting passengers. Rabbits, for example, may be nostalgically associated with the British countryside through tales like Beatrix Potter's, but they were brought from overseas. It may have been the Normans in the eleventh century, but according to one temporal window – a 2,000-year-old bunny tibia bone found in Chichester – Romans living in the UK also kept them as pets.[16] And from the 1500s onward, colonising Europeans began to routinely move flora and fauna back and forth around the globe. Corn and potatoes from South America were grown in Europe for the first time, and settlers in the Americas brought wheat, wine grapes, fruit and sugarcane. Scientists now know that this triggered forms of unnatural selection: when apple trees were introduced to the US, for example, some populations of native hawthorn flies started to feast on the fruit, rapidly evolving into a new species, now known as the apple maggot.[17] By the eighteenth century, the rise of global capitalism would supercharge the mixing, creating myriad new trade routes along which creatures could travel. The continents became an effective Pangea once more.

Towards the end of the 1700s, there was a further transition in humanity's relationship with nature, after people realised that huge stores of energy lay locked within ancient hydrocarbons. Through the burning of fossil fuels in the Industrial Revolution, wealthy nations began making significant alterations to the chemistry of atmosphere. Historians can describe how it happened as an economic story, but scientists can actually see it, written into the temporal windows of coral skeletons, ice cores, tree rings, stalagmites and ocean sediment. The factory emissions and pollution of this period also beget new species variants: melanism in peppered moths, and plants resistant to the chemicals in mine tailings.[18]

So, it's clear from the scientific timeview that human beings have always been influencing the living and physical world, something that's worth remembering before romanticising the nature of the recent past – sadly, the natural world has never been truly pristine while our species has been around. That said, science has also shown how the impacts have escalated – and from the mid-twentieth century, things kicked up another gear entirely, driven by a small fraction of the human population in developed countries. What has happened since then has been unprecedented in scale.

When future generations look back, tens of thousands of years from now, they will be able to see a temporal window in Earth's geology that represents a transition to a whole new planetary state. In the 1950s, there were so many atomic weapons tests that it doubled the amount of the radioisotope carbon-14 in the atmosphere. This 'bomb pulse' appears within glaciers, lakes, caves and reefs around the world. It's the now-official marker to represent where the Anthropocene began – a new geological era defined by humans.[19]

Scientists could have chosen various points in human history to mark the start of the Anthropocene – from early farming to the Industrial Revolution – but the geologists charged with making the decision selected the 1950s. Why? As well the bomb pulse, it was around this decade that the so-called 'Great Acceleration' began.[20] Alongside a sixfold increase in carbon emissions between 1950 and the present day, there have also been massive changes in biodiversity, pollution, ocean acidification, freshwater use, fisheries exploitation, tropical deforestation and much more. I needn't tell you about every one of these impacts, because you will know about them already, but here's a snapshot of some of the more startling changes . . .

The growth of the human footprint, for starters, has been staggering. In 1950, there were 2.5 billion people on the planet; now it's 7.8 billion. Today, half of the world's habitable space is used for agriculture, and there will be more urban areas built in the first three decades of this century than all of previous history combined. Meanwhile, the combined weight of inanimate human-made material – from tarmac to plastic – has grown so fast over the past 70–80 years that it now surpasses the Earth's total biomass. In other words, there's now more man-made material on the planet than every plant and tree, every animal, every microbe and every person alive right now. In the early twentieth century, it was only around 3 per cent. Roughly half of this anthropogenic mass is concrete, and if its rate of growth continues, it alone will outweigh the planet's biomass sometime around 2040.[21]

The Great Acceleration has also made our species into a greater evolutionary force than ever before. Many biologists fear we are causing the Earth's sixth mass extinction. And according to one estimate, the population sizes of wildlife have decreased by 60 per cent globally since 1970.[22] The proportion of mammals alone tells a story: in 1900, wild mammals made up 17 per cent of the land mammal biomass on the planet. Now it's only 2 per cent. The rest are livestock, pets, or us. Meanwhile, introduced or invasive species have gone into overdrive as the Pangea-like world we've created becomes ever-more connected: the number of wild Asian camels in Australia is doubling every decade, Mediterranean mussels have arrived in Antarctica and there's even a population of sub-Saharan hippos in South America, after they escaped from the druglord Pablo Escobar's zoo.[23]

Blistering evolutionary changes are now commonplace across the globe. Poaching during the Mozambican civil war during the late twentieth century, for instance, triggered the rapid evolution of tusklessness in African elephants, and trophy hunting in Canada has caused a 10 per cent decrease in horn size among bighorn sheep in less than twenty years. Remarkably, the average size of wild animals is, overall, smaller than it was a century ago. The body length in Chinook salmon in Alaska, for instance, has decreased by 5–7 per cent since the late 1970s, perhaps due to overexploitation (bigger fish get eaten; smaller ones are thrown back to pass on their genes), or maybe climate change (smaller bodies cope better with warming).[24]

We are now even changing geology itself. As well as the bomb pulse, there are beaches where the sand contains tiny fragments of Second World War shrapnel, new varieties of mineral leached from electronic waste dumps and, in Hawaii, a rock-like material called 'plastiglomerate': an intermingling of plastic, sediment, lava fragments and flotsam.[25]

So, while the scientific long view suggests that humanity has influenced nature from the start, it also shows that a massive experiment has begun that will transform the Earth and its life forms. Again, to be clear, this is an experiment in which some instigators carry far more blame and responsibility than others, but crucially, it is one in which we are all subjects. Nature is sometimes seen as separate from the human world, but our fate is entwined with the biosphere and Earth's physical systems. We rely on nature for food, fresh water, clean air, pollination, carbon sequestration, the mitigation of extreme weather, not to mention mental and physical well-being.

Can the Great Acceleration therefore continue? The researchers who are tracking it aren't certain. While it's possible that we could decouple growth from natural harms, human history is marked by myriad collapses preceded by climatic change, environmental degradation and rises in complexity – all of which are happening now. At the BBC in 2019, we published calculations by the researcher Luke Kemp from the University of Cambridge about how long civilisations and empires typically last: he calculated their lifespans across history and found an average of 336 years before things fell apart.[26] This suggests we don't have long to change trajectories – even more so if the clock started earlier than the 1950s, such as with the emergence of global capitalism in the eighteenth century.

There are some glimmers of hope – population is stabilising and there have been some environmental successes such as the restoration of the ozone layer – but it remains to be seen whether the future will bring a Great Decoupling, or a Great Collapse that brings our way of life crashing down as well as causing irreparable harms to other creatures.[27]

Towards the end of this chapter, we'll return to this question to look at some of the long-term strategies we might adopt to avoid the worst-case path, but first, let's return to our exploration of the scientific timeview, to make sure we have drawn all the long-minded lessons we can from its mode of thinking. Now that we've established the scale of the Anthropocene, let's take a few pages to explore one facet of the human impact on nature in greater depth – the discovery of global warming. This is a story worth recounting because it can tell us more about the way the scientific timeview works and how its long perspective gradually emerges. Temporal windows and models were one part of it, but there's more to it than that.

We could start it in many places, but let's begin in the nineteenth century. This was decades before it was established that humanity was warming the planet, but as we'll see, some long views do not appear overnight, nor do they come from a single person. But when they eventually do, the implications can be world-changing.

The discovery of global warming

In the mid-1800s, scientists identified a handful of temporal windows that proved difficult to explain. In the temperate latitudes of the US and Europe, geologists found curious features that didn't seem to belong there:

scratches on the surface of bedrock apparently gouged by huge volumes of ice, or giant boulders called 'erratics' that bore no resemblance to their surroundings in fields or forests. The only way to explain these geological puzzles was massive ice sheets, which must have once blanketed the land – and indeed entire swathes of the planet. This suggested the ancient Earth had once been plunged into ice ages. But how?

With no expert field of climate science, various researchers would promote their theories. Sun spots. Volcanoes. Shifts in ocean circulation. The gravitational pull of celestial bodies. When so much was unknown, all seemed possible.

Some of the first clues that fluctuations in carbon dioxide (CO_2) could have played a role can be traced back to 1856. An amateur US scientist and women's rights campaigner called Eunice Foote realised that CO_2 and water vapour trapped heat radiation. This was far from explaining the ice sheets, but with hindsight, it was the first demonstration that atmospheric gases could play a role in influencing climate.

However, due to Foote's gender and amateur status, she wouldn't get the credit in the history books for her observation. Her results were presented at scientific meetings – by male representatives – and reported in the US press, but she faced a subtle discrimination that meant her work only appeared in an obscure journal. Even the supportive male scientist who presented her results was gallantly condescending: 'The sphere of woman embraces not only the beautiful and the useful, but the true,' he told his peers.[28]

Foote's results made little impact in Europe, so recognition for the discovery of CO_2's importance in Earth's climate would instead go to the Irish scientist John Tyndall. In 1859, he independently staged laboratory experiments and came to similar conclusions as Foote. 'Thus the atmosphere admits of the entrance of the solar heat; but checks its exit, and the result is a tendency to accumulate heat at the surface of the planet,' he told a meeting of the UK Royal Institution. 'Such changes may in fact have produced all the mutations of climate which the researches of geologists reveal.'[29]

The next piece of the puzzle came in 1895, when the Swedish chemist Svante Arrhenius was approaching a divorce, and about to lose custody of his son. Perhaps as a distraction from the difficulties of his unhappy marriage, he decided to quantify the contribution of CO_2 to the greenhouse effect. With paper and pencil, he spent months calculating the atmospheric state and radiation for each latitudinal zone, to see if the gas could cause warming on a global scale. It was a massive job. 'I should certainly not have undertaken these tedious calculations if an extraordinary interest had not been associated with them,' he wrote.[30]

Arrhenius's calculations may not have been particularly precise compared to present-day measurements, but crucially, he confirmed that a global greenhouse effect was possible, and that rises and falls in atmospheric CO_2 could explain how the ancient climate had fluctuated so profoundly.

A couple of years later, in 1899, the geologist Thomas Chrowder Chamberlin wrote of a possible mechanism for the palaeoclimatic variations: he proposed volcanoes released large volumes of CO_2, which warmed the planet, which in turn accelerated CO_2 release from the oceans. Meanwhile, sedimentary rocks had the potential to lock up carbon in a calcareous prison, which could have led to cooling during periods of low volcanism. This would help to explain the ice sheets, he believed.[31] This wasn't quite right – there is a complex array of causes for the onset (and end) of ice ages, such as cyclical changes in the Earth's orbit around the Sun – but it remains true that falling atmospheric CO_2 levels are a major driver of cooling as well as warming. When long-term orbital shifts nudge extra carbon to begin to be locked up in oceans and vegetation, a global feedback effect begins, and temperatures start to drop.

What of human carbon emissions? Arrhenius was among the first to highlight that burning coal produces large volumes of carbon, and so could in theory change global temperatures in the future. He was not, however, troubled by the idea – it seemed many thousands of years away. Back then, the idea that human beings could play a major role in altering global temperatures was considered implausible. 'Hardly anyone imagined that human actions, so puny among the vast natural powers, could upset the balance that governed the planet as a whole,' writes the historian and physicist Spencer R. Weart, author of *The Discovery of Global Warming*. 'This view of nature – suprahuman, benevolent, and inherently stable – lay deep in most human cultures. It was traditionally tied up with a religious faith in the God-given order of the Universe, a flawless and imperturbable harmony.'[32]

Moreover, at the turn of the twentieth century other scientists began to (falsely) cast doubt on the idea that CO_2 could play such a major role in the natural climate. If the possibility of anthropogenic warming was already seen as unlikely, these claims rendered it apparently impossible.

It took decades for perceptions to begin to change, and even then it happened slowly. People had begun to notice that the winters in their part of the world seemed a little warmer than they remembered. Then in 1938, Guy Stewart Callendar, an engineer and amateur meteorologist, gave a talk to the Royal Meteorological Society in London, where he suggested that levels of CO_2 in the atmosphere appeared to be rising due to fossil fuels.[33] He wasn't warning of dangers – he believed it could be good for agriculture – but his calculations pointed to the truth. The meteorologists in the audience, however, were unconvinced. To the scientific establishment, Callendar was an outsider. And if anyone had cared to confirm his long view, the subsequent crisis of the Second World War would probably have made it difficult.

Nearly twenty years passed. Then, one day in the 1950s, the physicist Gilbert Plass was studying infrared radiation at John Hopkins University and came across the discredited theories about CO_2's role in past ice ages. His day job was weapons research, so he wrote about his investigations into the greenhouse effect in his evenings. Crucially, he was able to use early computers to help him calculate a more sophisticated simulation of the climate. Atmospheric models were hardly a new development – as far back as 1686, Edmond Halley had drawn circulating cells to explain the trade winds – but computation allowed for a far richer temporal picture than had been previously possible.

Using these techniques, alongside more accurate field and laboratory measurements, Plass concluded that humans had 'greatly disturbed the CO_2 balance' through fossil fuel burning and deforestation. Using his models to look to future possibilities, his calculations suggested that human activity could raise the average global temperatures at a rate of $1.1°C$ ($44°F$) per century, and warned that danger lay ahead for future generations: 'Even if there may be some question as to whether or not the general amelioration of the climate in the last 50 years has really been caused by increased industrial activity, there can be no doubt that this will become an increasingly serious problem as the level of industrial activity increases. In a few centuries the amount of carbon dioxide released into the atmosphere will have become so large that it will have a profound influence on our climate.'[34]

In the following decades, further scrutiny of the Earth's climate system lent ever-more insights into the long-term human warming of the planet. Other scientists showed it was far faster and more dangerous than a natural climatic swing, using the temporal windows of ice cores, tree rings and trapped pollen, while satellites and other instruments allowed them to observe and model transformations in weather, oceans and ice cover. All this allowed them to project ahead to the longer-term future: highlighting looming temperature increases that threatened global sea level rises, glacial collapses, ocean acidification, catastrophic wildfires and more.

In the late 1980s, this accumulation of evidence led to the formation of the Intergovernmental Panel on Climate Change (IPCC). Its goal since then has been to collate and review all the science into regular unified assessments of where we stand, and where we are headed. The IPCC would become one of the most long-minded organisations in the world: not only does it summarise a massive body of scientific evidence about the deep past, it has a long view that extends at least 100 years into the future too.

Thanks to all these scientists – and many more – we now have a clear view of the Earth's climate trajectory. Yet as our brief history of global warming's discovery shows, it took more than a century's worth of work to get here, featuring many setbacks and wrong turns along the way. What, then, might this tell us about the ingredients for long-mindedness, and how the scientific timeview emerges?

One of the most important lessons from the discovery of global warming is that this long view could only have been achieved *collectively*. There was no single insight or individual that unlocked the truth. If the story was a movie, there'd perhaps be one or two prescient, campaigning scientists seeing ahead to a future only to be ignored by their peers and society. The reality was far messier than that. It wasn't even as neat as I told it above – many hundreds more people were involved.

'The story of the discovery of global warming looks less like a processional march than like a scattering of groups wandering around an immense landscape,' writes Weart, the climate historian. 'Many of the scientists involved are scarcely aware of one another's existence. Over here we find a computer scientist calculating the flow of glaciers, over there an experimenter rotating a dishpan of water on a turntable, and off to the side a student with a needle teasing tiny shells out of a lump of mud.'[35]

Along the way, many of those scientists were unable to see the full picture, plenty were inaccurate or misguided, while others displayed the social biases and misconceptions of their time. Isaac Newton famously remarked: 'If I have seen further it is by standing on the shoulders of giants.' It's a piece of wisdom that is only partly accurate. While it's true that science can be cumulative, the reality is that it also often features wrong turns, competing paradigms and human fallibility.

Crucially, science's long view emerges – and endures – because it incorporates a diversity of theories and ideas. What makes it different from some of the other long perspectives that we've encountered in the book so far is that it does not rely on one belief system or ideology, apart from the scientific method. Yes, this is a worldview, and it's neither perfect nor comprehensive, but it works because it allows different paradigms to develop, and does not collapse if some of them are proved wrong. To draw on the words of Ludwig Wittgenstein: 'The strength of the thread does not reside in the fact that some one fibre runs through its whole length, but in the overlapping of many fibres.' Over time, some theories and observations show themselves as robust, while others fade away. 'Science gains strength, an ad hoc unity, from the fact that its key components rarely change together. Science maintains stability by being disunified,' writes the philosopher Sergio Sismondo, building on Wittgenstein's metaphor.[36]

A second long-minded lesson that might be derived from the story of global warming's discovery could be the length of time it took for it to be accepted and then to reach public and political consciousness. Living in the early twenty-first century, when human-triggered climate change is accepted as true by the majority of people – and 64 per cent of people in fifty countries believe it is an emergency – it's easy to forget that this was once much more of a minority perspective than it is today.[37]

Over the long term, societies can experience a form of forgetfulness about where once-heretical ideas have come from. It's an effect that psychologists call 'social cryptomnesia'.[38] Many of those who first raised the possibility of human warming were disbelieved, and plenty were seen as outsiders. Now that their core claims are accepted by all but the fringe, society as a whole has had a kind of amnesia about that fact.

With the benefit of hindsight, it's tempting to judge our predecessors for being slow to recognise the evidence for global warming, but I have to remind myself that if I had lived 100 years ago, I would probably have been in the sceptical or unwitting majority too. Following this logic, I must therefore be in at least one of those bubbles now, believing or doing something that my great-grandchildren may disapprove of – or at least see radically differently. The long view demands a form of humility about what we hold to be true and just.

That point leads us to a final question for this chapter, which takes us beyond the principles of the scientific timeview to reflect on its implications: when our descendants look at how we responded to climate change and the other transformations of the past century, what will they conclude? Science has given us a long view of our role within nature . . . so what should we now do with that knowledge? In short, what kind of Anthropocene do we want?

Benign fellowship and benevolent engineering

Given all the changes brought by global warming and the rest of the Great Acceleration, it might seem obvious what the solutions should be. If we are to reverse the trends of the past century, we will need to plot a path into the future where the effects of global warming are ameliorated, where the damage of our footprint is reduced, and where we avoid overexploiting our fellow animals and their ecosystems.

However, doing so will require difficult choices. It's unclear, at least to me, what the best long-term strategy should be if we want to live more harmoniously within nature while also balancing human needs. Should we aim for a kind of *benign fellowship*, restoring ecosystems to a wild state and drastically reducing humanity's footprint on Earth? Or should we aim for a form of *benevolent engineering*, where we proactively mould new ecosystems that permit human *and* animal flourishing? Are either of these approaches even possible? These are hard questions, not to be taken lightly.

Whatever path we pursue, there will be trade-offs. For benign fellowship, we'd need to step back from being the dominant species: allow space for rewilding to occur over large areas of the planet, so evolution could continue unimpeded, and reduce our greenhouse emissions to a pre-industrial state. This would be about living *with* and *alongside* nature, rather than in domination of it – a commitment to the self-determination of the natural world and the idea that non-human life has value in itself. This view shares ground with the perspective of 'deep ecology' – a term coined by the Norwegian philosopher and climber Arne Næss in the 1970s to describe a conservation approach based on 'the equal right to live and blossom' of all living things.[39]

In reality, this wouldn't be easy. For one thing, rewilding involves decisions about what kind of nature you wish to restore. Do you aim for a version as it was in prehistoric times, pre-industrial or pre-Anthropocene – or simply retreat entirely and hope life responds? Not all restorations are possible, and retreating from an already-unbalanced ecosystem could send it on a harmful trajectory.

More broadly, electing to become another variety of megafauna could require a severe curtailing of human population and potential. It makes sense to weigh up the needs of animals and people more fairly, but direct tensions are inevitable, and there are risks if this relationship became adversarial. Taking the side of nature over people on issues like population control can even lead to decidedly misanthropic conclusions and dark political paths.

But would benevolent engineering be any better? It also has both benefits and drawbacks. This is a view that sees the natural world as containing problems that could be solved by ingenuity and proactive intervention, improving animal welfare and prospects, and mitigating human impacts on living and physical systems. This might include conservation programmes in artificial human-managed spaces, or introducing species to fill ecological niches, but as technology develops, it could also feature projects like geoengineering the climate, or genetic interventions that save endangered species from extinction (or even bring lost ones back). The benevolent engineering approach is a view that does not see the natural world as sovereign territory that we should withdraw from, but as a space where we can use our advanced primate brains to manage the Earth's systems and help other species where we can.

However, a full-blown engineering approach could lead down problematic paths. The most obvious is that meddling with ecosystems and manipulating nature can beget changes with unanticipated consequences. Failed interventions in the past, like the 1930s introduction of the cane toad to Australia to control beetle populations, have caused irreparable damage to biodiversity. One misplaced instance of genetic engineering could destroy millions of years' worth of evolution, damaging the lives of many individual animals. And the idea of geoengineering the climate – such as injecting aerosols into the atmosphere – is fraught with risk, possibly making our climate predicament worse. Finally, there's a 'who are we to . . .?' objection that asks whether it is morally right for us to hold dominion over the living world, just because we have that power.

Hopefully in time, we can find a third way that balances the best of the two approaches. We need to, because continuing with the status quo is not an option. According to the official definition, we're only seventy years into Anthropocene, and already we've made irreversible changes, lost many species and set transformations in motion that will reach beyond the twenty-first century. If this new geological era continues for as long as the Holocene, it could last for another 10,000 years. Do we want the next 100 centuries to be defined the same way as this one? And can we realistically continue to flourish as a species if they are?

While the scientific timeview cannot alone tell us which strategies to embrace, it *can* show us where we stand before we do. Through its temporal windows, models and other perspectives, science has provided a richer, deeper understanding of life on this pale blue dot. Against the cosmic microwave background or the movement of the continents, we may seem inconsequential, but we are anything but. Ever since our species emerged, it has been entwined with nature. And now we have grown influential enough to shape our planet's entire trajectory. Ultimately, we need the scientific timeview if we want to reform this relationship with the natural world – only then can we figure out what kind of role a young species like ours ought to play.

We're now nearing the end of our tour of long-minded perspectives, but there's one more set of timeviews that we've yet to explore: the artistic, creative and symbolic approaches to imagining the long term. And it begins with a famous story about a grove of oaks, which has more to it than first appears . . .

The Persuasive Power
of Symbols and Stories

'A society grows great when old men plant trees whose shade they know they shall never sit in.'

ANON

'Art is certainly only a more direct version of reality.'

HENRI BERGSON[1]

Spend enough time talking to people about the long view, and somebody will tell you a story about a tree. They have become something of a symbol for long-term thinking.

Perhaps the most famous of all the stories is the one about the oaks of New College, Oxford. The tale goes that, sometime in the 1800s, officials realised they needed replacement beams for their main hall. To their surprise, they discovered that the college's founders had planted a grove of oaks in the 1300s to supply the job – an act of foresight and consideration for future generations.

Even the former British Prime Minister David Cameron recounted the story once during a Tory Party conference speech. 'Just think about that,' he continued. 'Centuries had passed . . . Columbus had reached

America ... Gravity had been discovered ... and when those oaks were needed, they were ready.' He then quoted his political predecessor: 'Margaret Thatcher once said: "We are in the business of planting trees for our children and grandchildren or we have no business being in politics at all."'[2]

Unfortunately for Cameron, the tale of the oaks is false. The land in Buckinghamshire, where the oaks supposedly came from, had not been acquired in the 1300s – it was bought later. 'I am amazed that this myth still continues: long-term tenacity if not long-term thinking,' the college archivist Jennifer Thorp once told me.

As we learned earlier, however, there is an example of arboreal long-sightedness elsewhere in the world that *is* true: the cypress trees planted a century ago to supply Japan's Shinto temples. Similarly, the UK Royal Navy once sought to plant trees to supply wood for their ships, knowing that it would take decades before they could be exploited. In the eighteenth century, there was a period of 'acorn fever', where British citizens were encouraged to plant trees for the Navy as an act of patriotism.[3] Supposedly, Vice Admiral Cuthbert Collingwood secretly dropped acorns from his breeches when he visited parks. (To this day, the British military still makes long-term plans for trees. It has a forestry team that manages almost 20,000 hectares of woodland – but now the goal is not exploitation of wood, but to increase its forestry footprint because trees are carbon sinks.[4])

Tree-based stories from fiction have also been deployed to inspire the long-term view. In Jean Giono's short story *The Man Who Planted Trees*, a young man encounters a shepherd living in a bare, desolate valley in the French Alps. Curious about why he'd choose to live in such a harsh place, the young man stays awhile, and observes the shepherd wandering miles to collect acorns. Years later, the young man returns, shell-shocked from the First World War, to discover saplings, which later become a verdant forest. By the end of the story, thousands of people live there, with little idea that they owe it all to a shepherd collecting acorns.

But my own favourite tree story is a bit more complex, and it goes like this.

One day in 1977, the artist David Nash travelled to a forest in Wales and planted twenty-two trees in a circle. His goal was to create a piece of art that would last long beyond his death, and so forge a tangible connection with longer-term time. That's not how it would turn out.

Nash was a sculptor living in Blaenau Ffestiniog, an ex-mining town where steep slopes of slate rise above the buildings like frozen, vertiginous waves. He looked at the economic and geopolitical turmoil in the wider world and decided he wanted to make a piece of artwork that transcended the era. 'The 1970s was a dangerous time,' Nash would later recall in *Christie's Magazine*.[5] 'People were saying that humans would destroy themselves before we got to the twenty-first century. So I thought: I'll make a sculpture that's aimed at the twenty-first century.'

First, he needed a location – he chose a clearing in a forest not far from his home in mountainous Snowdonia. After a few false starts where animals ate the saplings, he managed to get the twenty-two trees to take root in the circle. As they grew, he returned periodically to shape their branches so they entangled with one another as they reached upward. He called it *Ash Dome*. The artwork was an example of inter-arboration – the interweaving and overlapping of tree branches, a term coined by Thomas Browne in *The Garden of Cyrus*.

Over the years, Nash would produce various drawings of the dome as it grew, and many people became intrigued by the location of the project – but he kept that part a secret, and only a few people got to visit. However, his attempts to keep the dome safe from the outside world would be in vain. As the dome matured, it would face an unexpected threat.

In 1992, around 1,600km (1,000 miles) east in north-western Poland, researchers had noticed that there was something wrong with the trees there. The leaves had developed dark brown and orange lesions, and the shapes of diamonds on their stems. Eventually, the crown of the trees would wilt and disappear, and death shortly followed.

Closer analysis revealed it was a deadly fungus, thought to have travelled on exotic plants imported from Asia.[6] It began to kill ash trees – since only a small percentage were resistant – and spread fast across Europe.

A booming industry in the international trade of live plants and soil, the scientists realised, was accelerating its path across borders.[7] Imports and exports would eventually become more regulated, but it was too little, too late.

In 2012, the fungus arrived in the UK. By now, *Ash Dome* was thirty-five years old. What followed would be one of the worst tree epidemics the UK had ever seen, threatening 95 per cent of all ash trees in the country.[8] There was nothing that Nash could do to stop it reaching his natural sculpture. As the disease took hold and the trees began to die, Nash realised that the long-term legacy that he had hoped would outlive him could not be saved.

But that wasn't the end of the story. Nash, in his seventies, made a decision: he would plant oaks around the perimeter of the circle. As the ash trees die away, the oaks will reach up around them. In an act of inter-generational baton-passing, he also asked his son and grandchildren to help tend and shape the oaks after he dies. 'I can't make them look after it, but that is what's hoped for,' he told the *Guardian* in 2019.[9]

While the dome grows to be something quite different to what Nash originally envisaged in the 1970s, the idea will hopefully live on. And if so, it'll tell a richer story than he originally intended. If the fungus hadn't attacked the dome, Nash's project would have served as an effective symbolic artwork of long-term foresight, just like the other tree stories. But with the unexpected events of the decades that followed, it went on to stand for more. For one, it became a story of what happens when long-term plans rub up against the consequences of time-blinkered decisions: not Nash's, but others'. The disease was allowed to spread around the world in the first place because the desire for short-term profit in the plant trade combined with a short-sighted perception of risk.

Nash's story also adds another strand of evidence to support the idea that seeking to create a one-off legacy – the Patek Philippe strategy we encountered earlier (see page 191) – doesn't always go to plan. Embracing long-term thinking does not necessarily make you a good long-term fore-caster – even if you extend your gaze across the decades, unanticipated surprises can still come your way. But crucially, by adapting to the circumstances and recruiting his children to tend the artwork after his death, Nash did eventually find a way to transcend all those setbacks and discover a different form of long-mindedness – adaptive to change, and the unknowns of the future.

If the intergenerational care of *Ash Dome* continues, I'd like to think that the oaks will eventually be replaced by something else around their perimeter, and so on, as each layer approaches senescence. This isn't exactly what Nash himself intended, but it would be quite the legacy if the circle of trees were to get ever larger over time.

What is it about trees that makes them so effectively symbolic of long-term thinking? Plenty of non-human objects last beyond our lifetimes and leave a legacy for the future – a building, for instance. But perhaps the human relationship with the tree sparks the imagination because it is alive, and an organism with the potential to live longer than us – a vessel for our values to be carried forward. They also provide a helpful visual representation of time passing, via their branches or rings, and are accessible, familiar, local.

In a passage titled 'The Lessons of a Tree', the nineteenth-century poet Walt Whitman described the almost human qualities he saw in a 27-metre (90-foot) poplar in the forest near his home: 'How strong, vital, enduring! how dumbly eloquent! What suggestions of imperturbability and *being*, as against the human trait of mere *seeming*. Then the qualities, almost emotional, palpably artistic, heroic, of a tree; so innocent and harmless, yet so savage. It *is*, yet says nothing.' A tree, he acknowledged, cannot itself communicate wisdom, but nonetheless 'they do as well as most speaking, writing, poetry, sermons – or rather they do a great deal better'.[10]

As a symbol, a tree can encourage reflection about deeper tracts of time in ways that no mathematical calculation, policy paper or philosophical argument can. It's perhaps not surprising, then, that a number of writers, artists and organisations advocating for long-term thinking have converged on the tree as a symbol to represent their efforts: some have deployed tree rings as a logo, some have coined terms like 'acorn brain' to describe the long view, and some have embraced long-lived species like the bristlecone pine as unofficial mascots.[11] As the art historian Matthew Wilson observes: 'Great movements for change need powerful symbols. Picasso's dove is forever synonymous with the World Peace Council, and the rainbow flag inseparable from LGBTQ pride. These icons play a key role in unifying the actions of people from disparate backgrounds and nationalities. They bestow a visual identity on a set of ideals.'[12]

But what makes the symbolic nature of trees like the Oxford oaks or Nash's ashes so powerful is that they are also *stories*. When David Cameron used the story of the New College founders' foresight in a speech, it might have been apocryphal, but that didn't have much bearing on the intention: to inspire and to persuade.

Symbols stand in for abstract ideas and help us see more clearly, but narratives are how those concepts are circulated. If time is a currency that can be exchanged, its most valuable denomination would be the story. Stories are, in themselves, among the most powerful and enduring legacies a human being can leave behind for the world – far more so than a physical object.

Stories carry ideas, so that they can be told, repeated and embellished as the years pass. You don't need to visit the secret location of *Ash Dome* or the hall in New College Oxford to understand the messages embedded in those narratives; hearing about them is enough. Stories make the intractable tangible, and the incomprehensible accessible.

Even children can grasp extremely complex ideas through stories. When my daughter first encountered *The Very Hungry Caterpillar* as a toddler, she was learning more than the order of the numbers and the days as the caterpillar munches through increasing amounts of food during the course of a week. She was also being introduced to the idea that acts can have consequences when he inevitably ends up suffering a stomach ache, and perhaps even a dash of philosophy when he transforms into a butterfly, with the concept that the self you are in the present isn't necessarily who you always will be. *Winnie the Pooh* teaches friendship and gratitude, but also that it's OK not to have all the answers. *The Lorax* deals in rhyme but also corporate greed and conservation. And there have been multiple interpretations of the meaning underlying *The Tiger Who Came to Tea*, from allegories for 1960s feminism to the threat of authoritarianism.

As the cultural anthropologist Mary Catherine Bateson once observed: 'Wherever a story comes from, whether it is a familiar myth or a private memory, the retelling exemplifies the making of a connection from one pattern to another: a potential translation in which narrative becomes parable and the once upon a time comes to stand for some renascent truth. Our species thinks in metaphors and learns through stories.'[13]

Stories and symbols are also particularly helpful for moving the mind into the long-term future. When we imagine the past, it is richly painted. But the future is bare. To look ahead to humanity's trajectory over millions of years is to stare into a vast, empty space in which nothing is determinate. This asymmetry between past and future is a central problem for long-mindedness: how does one think about vast stretches of *bare time*? How do you persuade someone that the future matters when there's nothing there to see, or feel?

If the future is a vast, empty and unknown warehouse – a bare space of time – a story can be a painting on its wall. It doesn't reveal all the space's contours, and is just one possibility of many, but by filling nothing with something, it gives the mind something firmer to latch on to.

Stories have the power to transport us where we could not otherwise go, across time, space and into the minds of others – whether they are factual or fiction. As the author Kim Stanley Robinson has put it: 'You can go back to Rome while it burned. You can go to the moons of Jupiter, in the year 3000. You also, crucially, get telepathy: you get to know what someone else is thinking, and even experience it as if you were that person. It's a fictional experience, but nevertheless quite real ... The combination of what you know and have felt yourself with these new thoughts organised into characters can be like a real experience, or at least a waking dream. People who read fiction get to live 10,000 lives rather than one.'[14]

The power of analogy

But stories are not the only way to make the past and future more tangible. Another route is analogy. This can be particularly helpful for getting your head around the vast tracts of deep time.

Among the most famous deep-time analogies is one suggested by the writer John McPhee. While he may have been pessimistic about people's ability to imagine enormous numerical magnitudes, he did suggest that the landscape of deep time could be translated to physical dimensions we *can* understand: specifically, the distance between the tip of your nose and the tip of your outstretched arm (the old measure of the English yard). All of human history, he suggested, could be erased with the 'stroke of a medium-grained nail file' on the nail of your middle finger.[15]

But he wasn't the first to draw on a physical analogy for time. The writer Mark Twain also once dabbled with such metaphors. Twain had read Darwin and enjoyed hunting for fossils with a friend in the 1870s.[16] He even once made a contribution to the time-travel genre, writing the novel *A Connecticut Yankee in King Arthur's Court* in 1889 (which George Orwell memorably described as vulgar 'buffoonery'[17]).

His deep-time analogy transports the mind to Paris. In his essay 'Was the World Made for Man?', published fifty years after his death, Twain refers to 'man's share' of time as merely the 'skin of paint on the pinnacle-knob' at the summit of the Eiffel Tower. His wider goal was to skewer the anthropocentric hubris of his era, which claimed that millions of years of evolution had occurred only to prepare the world for us. He sketches out an evolutionary history from early invertebrates, through oysters (one of his favourite dishes), fish, dinosaurs, elks, sloths and kangaroos, until he reaches the monkey – and man. 'Man has been here 32,000 years. That it took 100 million years to prepare the world for him is proof that is what it was done for.' The skin of paint on the Eiffel Tower, he writes with tongue in cheek, was therefore the purpose of its construction.

In 1929, the physicist James Jeans would propose another evocative analogy, but for both the deep past and the deep future combined. He began by asking you to visualise a stamp laid on top of a penny, balanced on the tip of Cleopatra's needle in London, which is 21 metres (68.5 feet) tall. Working down from the top, Jeans proposed that the stamp represented the last 5,000 years of human civilisation, the penny and the stamp together was *Homo sapiens*, and the obelisk below was the total age of the Earth. But he also used this analogy to demonstrate how much more time could lay ahead for our species. Based on his (over)estimates of the total lifetime of the Sun, he imagined a trillion-year future for humanity. If the future lasted this long, he calculated that the stack of postage stamps would need to be as tall as Mont Blanc, which is 4,800 metres (15,780 feet) above sea level.[18]

Artists, of course, know all about the use of symbols and visual analogies. Over the past decade or so, a disparate collection of individuals and groups have attempted to foster a long-minded timeview. As we'll explore next, they have achieved this in several different ways; what unites them is their belief that art needs to inspire and engage the emotions and non-rational part of our brain.

Experiential and immersive

Taking a cue from the idea that spatial representations of time can be helpful, a number of artists and creative groups have sought to make time more accessible by converting it into analogical, often immersive, experiences.

The art installation *Donetsk Syndrome Diagrammatic*, for example, by the British 'geophilosopher' Paul Chaney, reduces deep time to the dimensions of a room. Its title refers to the town in Ukraine where conflict broke out with Russia in 2014, but Chaney leads up to that story by diagramatically depicting a timeline that begins with the Big Bang. Sheets of paper depict a deep-time timeline that snakes around the gallery walls, with elements and life emerging, tectonic plates shifting, all leading up to a regional war as if it were a tragic fate.

Other projects transport the mind to different places through immersion, rooted in the approach of *perspective taking* we first touched on in Part II. For example, recall how the design studio Superflux created an installation in the UAE that encouraged people to breathe the polluted air of 2020, 2028 and 2034. That's not their only attempt to transport people to other times experientially. Other Superflux projects – whose tagline is 'translating future uncertainty into present day choices' – have included the creation of the facsimile of an apartment in a world of severe climate change, and a collection of trees within a gallery intended to evoke 'a resurgent forest born from the ashes of humanity's hubris'.

In the first project, called *Mitigation of Shock*, the visitor could, as co-founder Anab Jain explained, wander around an 'apartment located in a future no one wants, but that may be on the horizon. Not to scare, or overwhelm, but to help people critically reflect upon their actions in the present, and introduce them to potential solutions for living in such a future.'[19] And in the second, the mini-forest called *Invocation for Hope*, the visitor could wander through burned and blackened pines – 'the unexpectedly graceful skeletons of a former time' – until they reach the heart, where life is returning around a freshwater pool. 'The installation leads viewers one by one on a personal journey from the ravages of climate crisis to the possibility of renewal and a deeper connection with nature,' the designers suggest.

Similarly, another perspective-taking art project, by Andy Merritt and Paul Smyth (known as Something & Son), imagines what the traces of a house from the local area might look like if 'fossilised' and then exposed at the surface within an imaginary archaeological site. Called *Future Fossil*, and based outdoors in Oxley Park, Milton Keynes, it is a negative cast of a building flecked with human-made materials. It may not be a scientifically accurate depiction of how a building would lithify, but the idea is, again, an exercise in immersive and empathetic time travel: transporting visitors to the perspective of some archaeologist in the deep future, looking back at the traces we have left behind from today.

Gifts to the future

If you could give a gift to future generations, what would it be? For the artist Katie Paterson, the message she has chosen to leave behind is a unique body of literature. Her project, called *Future Library*, began in 2014. Once a year, an author submits a manuscript to the Library that will not be read until the year 2114. Their books will be printed on paper made from 1,000 trees growing in a special forest called Nordmaka, near Oslo in Norway. As of 2021, there were eight authors signed up, including Tsitsi Dangarembga, Ocean Vuong, Karl Ove Knausgård, Han Kang, Sjón and Elif Shafak. Margaret Atwood contributed a story called *Scribbler Moon*, while David Mitchell's is called *From Me Flows What You Call Time*. All the stories will be kept secret for a century, so only our grandchildren will get to read them.

Paterson's project is one of many that might collectively be described as 'gifts to the future'. Another example is the legacy that Jonathon Keats intends to leave behind: a photograph that he himself will never see in its finished form. In 2015, he placed a simple pinhole camera in Stearns Steeple, a church in Tempe, Arizona, that will take an exposure of the town that will reveal itself gradually over 1,000 years. 'I'll be dead,' Keats said at the project's launch. 'But I don't regret it at all. For me, it's much more interesting to be here today, seeing the behaviour of people who know they're being watched by the unborn, and also to be watched myself, living vicariously as a future memory of the next millennium.'

Some future-gift projects are more whimsical. For instance, I own a copy of an invitation to a party in 2269 that only my descendants can attend. Designed by Peter Dean and Michael Ogden, it comes in the form of a poster print that is intended to be passed down to my grandchildren. The pair were inspired by Stephen Hawking's 'party for time-travellers'. Hawking once threw a champagne and cocktails event for people from the future – he only sent out the invites afterward (no one came).

Others emerge from grander aspirations. The *Memory of Mankind* project, created by Martin Kunze, is a time capsule for knowledge. It bears striking similarities with the sci-fi writer Isaac Asimov's famous 'Foundation', which he imagined as a body of all information, left for those that survive the fall of civilisation.

By collaborating with universities, libraries and newspapers, Kunze is creating a record of documents, papers, novels, news stories and other details about our culture to be buried in a salt mine in Hallstatt, Austria. Rather than relying on digital storage, which is ephemeral and vulnerable to bit-rot or medium obsolescence, Kunze etches information on to what he calls 'ceramic microfilm'. He covers plates with a dark microfilm before using a laser to write on to them. Each one can hold five million characters, and they are extremely durable. It is an eclectic and subjective record, featuring local history, scientific patents, everyday objects like washing machines, the sites of nuclear waste dumps – and even lengthy descriptions of celebrities like David Hasselhoff.[20] But the idea is that, should a future society need it, the information will all be there to guide them.

Like all time capsules, these projects do have some limitations: they represent what we would *like* to leave behind, rather than what future generations might actually *want* from us. They are a subjective snapshot of what we value in the early twenty-first century. But the point is to encourage that reflection in the first place.

They are asking people to think about the asymmetry between the present and future. In the case of Paterson's *Future Library*, which prevents people from reading the works of admired, commercially successful authors – imagine the sales figures for Atwood's *Scribbler Moon* if it was published today – Paterson is drawing attention to the rarity of such a selfless act. It's not often we gift something to future generations that has no benefit for our present-day selves. Compare it with a legacy like the presidential library: while it's ostensibly a gift, the motivation is usually to enshrine the memory and glory of the politician that donates their name.

Long-lasting and slow

Another category of symbolic project encourages a different form of reflection by leaving a different form of legacy: these are projects that aim to capture the slow pace of deeper time, connecting present-day people with the unborn via shared experiences that last thousands of years.

Perhaps the most famous of these 'slow time' projects was conceived in the late 1990s, when Jem Finer, one of the founding members of The Pogues, decided he wanted to make a piece of music that would last for more than 1,000 years. As a child, he had looked at the stars and marvelled at the idea that their light had taken millions of years to reach him. Now he was older, he was drawn again to the vastness of time, and also how, when playing music, the perception of it felt different, almost controllable.[21]

So on 1 January 2000, he started *Longplayer*, a musical score designed to play without repetition until the last day of 2999, before repeating again. It's an eerie, but calming, composition seemingly intended to evoke a feeling of religiosity in its listeners. The installation can be heard at a lighthouse in London, with 234 Tibetan singing bowls used in live concerts to accompany the score, but there's also an online livestream at longplayer.org.

As *Longplayer* trustee Gavin Starks writes: '*Longplayer* helps us ask many questions about our world and our role in its future. It helps frame questions that are much bigger than us – but they are not "infinite". The time-bound nature of the project leads to many different questions: what might be happening in the future? What might our role be? What might our impact be? How might we communicate across 40 generations? What will be happening on its fifth thousand-year loop?'

A few months after Finer's *Longplayer* composition began, another long-term piece of music also started playing, elsewhere in Europe. In the 1980s, the composer John Cage had written a piece called *As*

Slow as Possible. He died in 1992, but at the turn of the millennium the John Cage Organ Foundation decided to take his instructions to play as slow as possible to heart. In a 1,000-year-old church in Halberstadt, Germany, they built an organ that will take 639 years to play Cage's full composition. Powered by bellows, the organ plays a single note for months at a time.

There must have been something in the air in Germany in the 1990s, because it's not the only project the country's creatives conceived during this period that deals with the theme of time's long and slow passage. In 1996, Bogomir Ecker created an artwork called *Die Tropfsteinmachine*, in which an artificial stalactite and stalagmite are forming in the Hamburger Kunsthaller gallery. Ecker designed it to run for 500 years. Rainwater enters from the roof, before seeping through plants and earth on the ground floor until it reaches the machine in the basement. A steady series of drops fall on to a stone slab, precipitating minerals as rock at a rate of 10mm (½in) every 100 years.[22]

Also in Germany is the public installation *Zeitpyramide*, conceived by Manfred Laber. Begun in 1993, on the 1,200th anniversary of its host town Wemding, it is intended to be a giant pyramid built from cubic concrete blocks that will take more than 1,000 years to be completed, in the year 3183. As of 2021, the first three of its 120 concrete blocks have been placed. While it may look simple now, the idea is that future generations will see it in its full form.

Like the Pitch Drop Experiment we encountered in Part II, all of these artworks demonstrate in a creative form – musically, tangibly, artistically – that much of the world passes at a pace imperceptible to the senses, and that our acts today can reach further into the future than we might realise.

Will they all last as long as their creators intend? Perhaps some are more likely than others. *Zeitpyramide*, for example, relies on the authorities in Wemding to carry on funding its construction for a millennium, which is far from certain. And *Die Tropfsteinmachine* will be on public view in the Hamburg gallery only for as long as its curators want it there.

The others may have a slightly better chance. The organ that plays *As Slow as Possible* may well endure because it is situated within a church, and because of the long-lived religion attached to it. Also, Cage's score itself will always be available for anyone who wants to play it.

Longplayer, meanwhile, is overseen by a trust established to last as long as the project. That said, there is a second piece of music by Finer that potentially has just as strong a chance of enduring across the long term. In the mid-1980s, Finer penned a folksy ballad with The Pogues about a sailor in New York dreaming of home in Ireland. Dissatisfied with the lyrics, the band decided to replace them with a boozy nostalgic duet, set at Christmas. They released it in 1987 and called it 'Fairytale of New York'. A long-minded friend of mine once suggested to me that this song could endure just as long as *Longplayer*, because it is repeated and remembered by millions of people as a tradition every Christmas. (In the UK, it is the most-played festive track of the twenty-first century so far.[23]) As we've learned, one of the greatest sources of longevity is repetition and ritual. And that leads us on to the next category of symbolic creative projects: those that invite intergenerational participation.

Participatory art

Around 3,000 years ago, an ancient community living in what is now Oxfordshire in the UK decided to create a giant work of art. It's not known why, but perhaps it was to please their gods, because it would be directed at the sky.

The people climbed up a small hillside, and began to dig and carve trenches, which they filled with crushed chalk. Their efforts stretched right across the hillside. Later, it would be named the White Horse of Uffington. At the time, it wouldn't have been possible to see it from the air, but now we can, it looks like this:

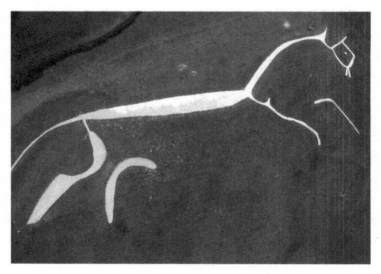

The White Horse of Uffington from the air.

Notably, the White Horse has endured far longer than almost anything else these Bronze Age people left behind. Why so? Because it has been continually maintained ever since.

In the summer of 2021, I found myself crouched over the surface of the horse along with several others, bashing rocky chunks of chalk into a fine powder. Wearing gloves and wielding hammers, we spent the afternoon 'chalking the horse' on a day out organised by the architect and designer Chris Daniel and guided by a team from the National Trust who oversee it.

Depending on how seriously you take yourself, participating in the horse's annual re-chalking is either a secular ritual that allows one to connect mindfully with past and future generations – or it's a fun day out to the countryside. When I told my wife I was going to chalk the horse, she joked that it would be just me and twenty other authors who aspired to write about the experience in their books, each of us remarking on *the ancient whispers of lost ways* or *the humbling purity of working with one's hands*. Speaking for myself, she wasn't far wrong because I'm writing about it now, but mainly it was simply enjoyable. For me, it also cemented the importance of why long-mindedness is at its most accessible when it is communal. People who previously didn't know each other were brought together to spend the day participating in a long-term act, and so naturally many of the conversations tended to be about the deeper past and future as we worked.

The activity also spoke to the symbolic value of *upkeep*. In recent years, a growing number of people have begun to argue that the act of maintenance tends to be an undervalued by present-day societies, despite being at the root of the longest-term endeavours. Whereas it is often the inventors that win the status and plaudits in twenty-first-century life, it is the maintainers in a society who make things *last*, according to the science-and-technology-studies scholars Andrew L. Russell and Lee Vinsel.[24] 'Western cultures laud and heap attention on so-called "innovators" but often overlook and forsake those individuals who keep our world from falling apart,' they say.[25] 'These crucial individuals keep society's systems running.' Maintenance can be physical – maintaining infrastructure, bridges, roads and so on – but also non-physical too, via the continuation of long-lived ideas, institutions or practices.

True to their words, Russell and Vinsel have co-founded an organisation with the goal of celebrating and studying acts of upkeep, called The Maintainers. It actually started with a joke, but then got serious. Walter Isaacson's book *The Innovators: How a Group of Hackers, Geniuses, and Geeks Created the Digital Revolution* had just been published, so they joked about writing a book in response called *The Maintainers: How Bureaucrats, Standards Engineers, and Introverts Created Technologies that Kind of Work Most of the Time*. 'We had a good laugh – but then we shared the joke with some of our friends, and the idea took on a life of its own,' the pair recall.

But participatory acts of continuation do not all need to focus on the preservation of what already exists. It is possible to do both: to come up with a great idea today, and invite participation that extends into the future. An excellent example of this can be found in the Dutch city of Utrecht. There, a poem is being written over many years, letter by letter, with different poets passing the baton between one another. Every Saturday at lunchtime, a stonemason in the city centre carves a new letter into a cubic block. He or she then places the new letter next to the last one, embedded in the street. It's called *The Letters of Utrecht*, and the poem is currently about 100 metres (328 feet) long.

It begins with the words:

Je zult ergens moeten beginnen om het verleden een plaats te geven, het heden doet er steeds minder toe. Hoe verder je bent, hoe beter. Ga maar door nu, laat je sporen na.

Which roughly translates as:

You have to begin somewhere to give the past its place, the present matters ever less. The further you are, the better. Continue now, leave your footprints.

What makes the project so impactful is that every week, the number of people invested in the project and its continuation grows. When the stonemasons are finished, members of the public sometimes scrawl their names on to the sides with pen or correction fluid before it is laid into the ground. For €100, it's also possible to donate a stone and attach a name permanently to the record by having it carved onto the side.

Like other works of art, it visualises the passage of time, but is fully public – placed centrally within a town as people go about their business, and all the while emphasising long-minded values of baton-passing and collectivism. 'The poem continues for as long as there are Saturdays. Until the end of the future,' its overseers write. 'The poem is never finished. A work of art for future generations.'

The *Letters of Utrecht* project helps to connect everyone in the city – but also, in a way, reaches beyond to link up with communities much farther afield. How so? The very first stone in the poem – carved with the letter J – was donated by an organisation thousands of miles away: the Long Now Foundation, in San Francisco, which has the long view at its heart.

Their donated stone comes all the way from the US, where the foundation is building its own ambitious work of temporal symbolism.

In fact, Long Now's project could well be one of the grandest symbols for long-term time that our generation will create: a clock that will tick for 10,000 years. It combines immersion, a gift to the future, long-lived music and even a dash of participation – all deep within a mountain in Texas. Dreamed up by a group of Silicon Valley visionaries, computer scientists and a pioneering musician, and backed by one of the world's richest people, it even has its own origin story, beginning in a bad neighbourhood in the late 1970s.

Grand symbols

The music producer Brian Eno was in a run-down corner of New York, on the way to a glamorous dinner party. It was the winter of 1978, and Eno's taxi was bumping over potholes, hurtling towards an address he didn't recognise. As he was driven south, the streets got darker and the sense of urban neglect grew, until finally he arrived at his destination. A man lay slumped in the doorway. Puzzled, Eno double-checked the address on the invitation card. He had been invited to the home of a wealthy public figure for dinner. Could this really be the right place?

Eno rang the bell and rode the elevator up to the apartment. Inside, to his surprise, was a glittering, glamorous loft probably worth $2–3 million. Curious, he asked the hostess during dinner if she liked living where she did. 'Oh sure,' she replied, 'this is the loveliest place I have ever lived.'

He realised that what she meant was 'within these four walls'. The dilapidated neighbourhood outside didn't exist for her. Afterward, when he looked around at his contemporaries, Eno saw the host's narrow view everywhere. What's more, this attitude to space also translated to the way this New York glitterati thought about time – not much further than the following week. They were living in what Eno called a 'small here' and a 'short now'. 'Everything was exciting, fast, current, and temporary.

Enormous buildings came and went, careers rose and crashed in weeks. You rarely got the feeling that anyone had the time to think two years ahead, let alone ten or a hundred,' he later reflected.[26] 'More and more,' he would write in his notebook, 'I find I want to be living in a Big Here and a Long Now.'

Years later, this experience inspired Eno to work with several other like-minded thinkers, including the technologists Stewart Brand and Danny Hillis, to co-found the Long Now Foundation, which aims to 'provide a counterpoint to today's accelerating culture and help make long-term thinking more common'.[27] It runs regular speaker events in San Francisco, and has launched initiatives such as the Rosetta Project, a digital library of all human languages designed to last millennia, and a website called Long Bets, which asks people to stake money on their predictions for the long-term future. They also have offshoots in Boston, London and Barcelona. The chalking of the White Horse of Uffington was organised by the foundation's British chapter, led by the architect Chris Daniel.

Their scope of view is 10,000 years hence, because it was around ten millennia ago that agriculture became widespread, and civilisations began. When talking about dates, they prefix them with an extra 0 in order to capture the notion that our paltry years are dwarfed by a larger timeframe. So for a Long Now follower, we're now in the 02020s. Perhaps unsurprisingly, they also have a strong affection for symbolic trees, particularly the bristlecone pine, which is their unofficial mascot. (These are some of the oldest trees in the world. A bristlecone called Methuselah in eastern California is more than 4,850 years old.)

Of all the Long Now Foundation's projects, however, the most ambitious attempt to embrace the long view is its 60-metre (200-feet)-tall 10,000 Year Clock, currently being installed in the mountains of West Texas on land owned by the Amazon billionaire Jeff Bezos. The Clock is intended to be a monument that outlasts its creators, ticking for ten millennia. The first parts of the Clock's mechanical architecture have now been placed within the limestone cavern, but it's been a project decades in the making.

Inside the mountain, a 16-tonne diamond-chainsaw robot has carved a spiral rock staircase, which will snake around the metal cogs and gears of the Clock in a central cavity hundreds of feet deep. Engineers have installed a manual winding mechanism to power the bells and display dials, but the Clock itself will be kept running by the temperature difference from day to night. Air within a tank and bellows at the cavern's top will expand during the day, providing just enough energy to keep a pendulum ticking for millennia.

As the centuries pass, a new, different sequence of bells will play every so often, a long-lived auditory experience like *Longplayer* or *As Slow as Possible*. You can get a sense of what future generations might hear on one of Eno's albums, inspired by the Clock, called *January 07003 Bell Studies*. The first track features the sequence of bells that will play 5,000 years from the year he composed it – in the year 07003.

When it is finished, the hope is to provoke the Clock's visitors to reflect on their place in time – standing amid the ancient geology of a cavern looking at a mechanism that will tick for hundreds of lifetimes. And while it is designed to continue without human intervention, tuned to account for changes in the Earth's rotational pace and wobble over millennia, it does encourage some participatory involvement, allowing future people to help it display time by physically winding it via levers.

The building of it, in itself, has been a lesson in how to think longer term. Long Now's Alexander Rose tells me they have had to contend with a variety of future events that could destroy their project, which a typical architect or engineer doesn't have to worry about.[28] For starters, over thousands of years, you have to plan for long-term climatic or geological shifts such as sea level rise or earthquakes. Many ancient constructions have been lost to rising waters over the centuries. Then there's the worry of ideological change. Political decisions can also rapidly destroy architectural legacies following shifts in beliefs. For example, in 2001 the Taliban blew up the Buddhas of Bamiyan in Afghanistan, giant cliff statues built in the sixth century. A legacy can also be dismantled when a society's values evolve to reassess its past: consider how Black Lives Matter protestors pulled down statues of once-venerated businessmen and politicians attached to the slave trade. The Clock might not seem so controversial, but the point is that values that are accepted by our generation can change over very long-term time.

Finally, across time, all materials decay. The lifespan of brick and concrete is more than a century, but a building's life expectancy can drop as low as sixty years if its architect relies on steel, glass and synthetic materials. Its actual expectancy also depends on location, influenced by both climate and social surroundings. One study of 1,732 buildings in Chongqing found an average lifespan of only thirty-four years, which was shorter than they were designed for.[29] On the timescale of millennia, even hard-wearing materials like stone decay eventually. The ancient Parthenon of Athens, or the abandoned priories and castles of the UK, are still here – but they are ruins.

As a result, the Clock includes design features like ceramic ball-bearings that are far more resistant to wear and tear than metal versions, as well as replaceable parts that allow for ongoing maintenance by future generations. Rose acknowledges that it's not possible to account for ideological or political shifts that might seek to abandon or tear it down, but its remote location may at least help avoid that fate.

Myself, I have mixed feelings about the Clock. I wonder what future people will make of it, looking back at the period and the place in which it was conceived. It is perhaps more of a legacy project than an intergenerational endeavour like *Ash Dome*, the White Horse or *The Letters of Utrecht*, and therefore I fear for its future once the original Long Now founders have long gone. Ironically, it is already emblematic of a period that has now passed, beginning its development amidst the dizzying and optimistic early years of Silicon Valley growth at the turn of the twenty-first century. Nowadays the attitude that permeated those heady days is cynically branded as techno-utopian, standing in sharp contrast to the decidedly more flawed digital world that Silicon Valley has actually created.

Critics of the Clock also like to take potshots at its funder Bezos. He isn't actually involved in the day-to-day operations of Long Now, which is a small, lean organisation that relies on donations to keep going, but the cynics have suggested that Bezos's riches might be better spent on taxes, problems like climate change, or social programmes that benefit present and future people. When I think of the Greta Thunberg generation, with their values of planetary stewardship and suspicion of industrial capitalism, it's difficult to see why they would prioritise the upkeep of a symbol that emerged during this specific period of American largesse.

While it's difficult to say what future generations will make of the Clock, I do hope, however, that they can see it as its makers intend: a symbol for long-term thinking and an attempt to highlight modern society's time-blinkered predicament. To their credit, those now running the Long Now Foundation are well aware that the Clock is just one of many routes to long-mindedness. In recent years, they have been exploring myriad other plans for their 'second quarter' (their *next* twenty-five years), including hopes to support and foster ideas and projects more globally, and a greater emphasis on diversity.

Over 10,000 years, the associations with polarising figures like Jeff Bezos and places like Silicon Valley will probably be forgotten anyway, or at least hidden in a history book. Perhaps the Clock will mean something entirely different to our descendants, revealing a truth of our age that we can't yet imagine: a symbol that will go on to say more about what we value and who we are today than we will ever know ourselves. If it does hold a hidden truth about us that only our descendants will see, I suppose it would be an appropriate legacy for the long-forgotten billionaire and founders that placed it inside a mountain.

Before we bring our tour of long-minded timeviews to a close, let's conclude with one more story about a tree.

It turns out that *Ash Dome* would not be the only project of David Nash's that would speak to the human relationship with time. Nor was it the only one with unanticipated consequences. Another major work of his was *Wooden Boulder*, whose story – which I owe to the art historian James Fox – began a few months after he planted the ash trees.[30]

One day in the winter of 1977, Nash learned that a giant 200-year-old oak tree had fallen in a storm high up in a valley not far from where he lived in Wales. He was short of money at the time, and he needed wood to make sculptures in his studio. So, he trekked up the hill, and carved a boulder-shaped chunk from the fallen trunk. Around a metre in diameter, it had edges and facets like an enormous semi-spherical crystal.

He soon realised, however, that the boulder would be too heavy to carry back to his studio. So, he pushed it into the nearby Bronturnor stream, in the hope that the water would carry it down the hill. It went fine at first but to his dismay, the boulder soon got stuck between the stream's rocks. All he could do was go home, and wait.

Six months later, heavy rainfall dislodged it, allowing him to retrieve it from the pool. But a few days later, he learned some teenagers had pushed it back in. Once again, he had to drag it out. He left it at the top of a waterfall a bit further down.

Around this point, Nash decided to change his plan. Perhaps, he thought, the goal should not be to get the boulder home any more, but to let it take whatever path it wanted. He thought of it as a 'free-range' sculpture.

Over many years, it moved a little bit further down-river. Sometimes, heavy storms would push it on; sometimes, Nash intervened to keep the boulder's story going. Occasionally, the local council were on the verge of removing and destroying it, unaware it had artistic worth. Sometimes, it would disappear altogether, only for Nash to learn of its new location through a phone call.

After 2003, it was only sighted once over a ten-year span, until it emerged again in summer 2013 in a Welsh estuary, more than 50km (31 miles) from its birthplace. Nash jumped in a car to go and look at it one last time. After that, it was gone. Perhaps it finally came to rest and is slowly being buried beneath sediment, or maybe it's still drifting out there somewhere in the ocean.

'*Wooden Boulder* underpins everything I do. It's where I've really got it,' Nash told Fox in *Christie's Magazine*. 'It's probably my most satisfactory statement as an artist.'

It's also one of my own favourite symbols for the human relationship with time. Like *Ash Dome*, the story illustrates that the future brings setbacks, no matter how well you plan. *Wooden Boulder*, though, also speaks to something else: the truth that we are all navigating a river towards the open sea. But while disappearance and oblivion may lie ahead eventually, that's for tomorrow: along the way, there is meaning to be found.

12

A Deep Civilisation

'What are we of this generation, or of any other generation, that we should usurp, and expend upon our particular and exclusive uses, what was meant for mankind? It is lent to us, not given: and it is our duty to pass it on, not merely undiminished, but with interest.'

JOHN STUART MILL[1]

As I've watched my daughter Grace grow up, she has developed an ever-greater awareness of time, and how she fits into it. Her perspective is changing every year – and witnessing that development has informed my own.

When she was three, she had no sense of clocks or calendars. She could understand the basic events of *The Very Hungry Caterpillar*, the classic children's book about a creature gorging food over a week, but when she told the story back to me, she would get confused about the order of the days. Time, for her, had no structure.

By the age of five, that had changed, and she had figured out how yesterday trailed behind her and tomorrow extended ahead. At breakfast one day, I asked her if she knew what the future was.

She paused. 'No, not really.'

'Well, you know history, and the past? This is the opposite.'

She chewed her cereal.

'What's the furthest in the future you can imagine?' I asked.

'Um . . . when I am ten.'

'Can you imagine further? Being a grown-up?'

'No. When I am ten.'

She picked up her bowl and wandered out to the kitchen. Tomorrow existed for her, I realised, but went dark a few years ahead.

When she was seven years old, we had another conversation at the dinner table about time. I asked her how much she thinks about the future.

'Not often,' she said. 'But sometimes I worry what will happen.'

'What do you worry about?'

'Getting hurt or getting arrested or something.'

'Can you imagine being the same age as me and Mum?'

'No.'

'Can you imagine being a teenager?'

'Yes.'

'Can you imagine having your own children?'

'That freaks me out.'

Aged eight, history and geology caught her interest. She'd come home from school to tell me about the Egyptians or the Tudors, and started collecting a small menagerie of fossils. She also began to populate her future with hypothetical events. The media and culture she encountered tended to fill up that canvas. I'd often no idea where she picked it up: sci-fi visions like time travel, or the idea of living in space.

'The Singulation,' she explained to me, 'is where people are miserable in the future. And a person says "what's the point?" The robots take over the Earth.'

'Wait, are you talking about the Singularity?' I asked. Somehow, she had picked up on the transhumanist theory that computers will soon overtake human intelligence. 'Where on earth did you learn that?'

A cartoon, she replied. *Captain Underpants.*

*   *   *

When I reflect on these conversations, I am reminded that we are not born with a long view – we come to discover it. And along the way, it will be shaped by cultural influences, other people and the surrounding world.

When I first set out to write this book, many years ago, I confess my interest in cultivating a deeper temporal perspective was heavily shaped by my fears: the threats of tomorrow, and a concern that short-termism was hiding dangers ahead. I imagined my daughter's trajectory to the twenty-second century, and saw a future world marred by climate change, environmental degradation, technological missteps or worse. I still worry deeply about the mistakes our generation is making, and the malignant heirlooms we are leaving behind. And I have no illusions about the fact that we are living in a difficult and precarious period. We need the long view to redirect our trajectory and avoid perils that could be far worse. But I have since realised that a long-term perspective offers so much more: it can also lead to a more hopeful view of the world, where the prospect of positive change feels ever-more possible.

Here's what I have learned on this journey about the many upsides of the long view; about what it means and what it doesn't, and about the ways everyone would benefit if they embraced it:

THE LONG VIEW IS RESTORATIVE

It can be tempting to assume that the long view is an exercise in sacrifice: a solemn, weighty duty that involves forgoing present-day pleasures. However, there are many personal advantages to embracing a deeper view of time. I've learned that it offers perspective amid upheaval and a source of energy and autonomy when the bad news feels overwhelming. In times of crisis, every step forward is hard and every setback can make oblivion feel closer, but amid such struggles, the long view offers fortitude and perspective.

THE LONG VIEW IS A WAYFINDER

If we want to chart a route out of difficult times – steering our fate rather than stumbling into the future – we need a compass direction. This is what the long view provides: a form of guidance for navigating a complex world. But as well as offering a route around future dangers, it also reveals the learnings of the paths already taken – alternative histories that could have been – and that there are myriad trajectories that could lie ahead. To be long-minded is to know that there are always multiple possibilities and turning points as we move through time. The long view reveals that the future is only singular when it becomes the present; until that day, it is always *plural*.

THE LONG VIEW MAKES THE PRESENT MORE MEANINGFUL

A misconception about long-term thinking is that it involves spending all of one's mental time in the past or future, detached from the world as it is now. On the contrary, I believe that looking at the world through a temporal lens lends greater meaning to life in the present. Discarding blinkered short-termism allows you to become more *present-minded*: able to see with far clearer sight what truly matters, what needs to change, what is dangerous and harmful – and what is worth enjoying and appreciating in the world today.

The long view also provides a clarity of purpose in the present: the duty to posterity. Some might interpret that this means building a grand legacy. However, I do not believe that we need to use all our energies fashioning heirlooms for our descendants, nor do we need to solve all their problems, or impart our wills on the future. After all, we can't predict their needs and values any more than someone living 1,000 years ago could imagine ours today. The greatest legacy we can leave behind is simply *choice*. If we can ensure that tomorrow's people have the means and autonomy to decide their own path within a sustainable world, then that is enough.

THE LONG VIEW CAN BE ACCESSIBLE TO EVERYONE

I am aware that not everyone has the stability or comforts to think beyond the difficulties of their present – many people need help and support, rather than lectures in long-mindedness. That said, I believe that a long-term perspective needn't require deep pockets or plentiful resources. Nor should it be seen as a predilection for the elite. Cultivating it can start locally, within a family or community: a conversation with a friend or loved one about the past or future, ancestors and descendants, or by seeking out the signatures of deeper time within a neighbourhood or surroundings. The long view needn't be a luxury, it can be part of everyday life.

THE LONG VIEW IS DEMOCRATIC

Over the years, I have encountered many different timeviews – religious, indigenous, philosophical, scientific and creative. History also shows that past cultures had their own temporal perspectives, shaped by their knowledge, beliefs and assumptions at the time. Often these long views speak different languages, with different priorities and values: some are transcendental and rooted in faith; others are secular and empirical. Some span timescales of centuries; others millions of years. Some focus purely on humanity; others take in the natural world too. But this is OK. The long view belongs to everyone, and it should be a democratic, collective enterprise. It would be a homogenously dull world if everyone embraced the same form of long-mindedness.

In the coming years, I hope whole new timeviews soon emerge. There are people, professions and communities who have yet to fully discover the potential of the long view, but when they do, they will no doubt build deeper temporal perspectives in their own way. Future generations will do the same: there may be insights about time and the world that are yet to be discovered, which lead to forms of long-mindedness that I cannot imagine.

THE LONG VIEW CAN BE POLITICALLY UNIFYING

While there will always be multiple timeviews, the core principle that *the long term matters* is usually something that people of all perspectives can agree on. After all, short-termism has few defenders, and over the decades, politicians of all stripes have referenced the duty to posterity in their speeches – left and right, liberal and conservative. In an age of political polarisation, the long view can therefore remind people of their shared values: whether it is the importance of tradition, ancestry and the lessons of history, or the belief in handing our children a better world.

THE LONG VIEW LEADS TO A HEALTHIER MEDIA DIET

Every day, we are exposed to a barrage of noisy information, making the forces and changes that matter harder to see. 'Salient' does not necessarily mean 'important'. Sometimes, major news events do affect the long term – wars, pandemics, elections. However, so much else of the daily churn is ephemeral, and even unhelpful, for understanding how the world truly works. It's the information equivalent of junk food.

Our mental models of the world are shaped by the information going in. So, if I spend all my time reading nothing but the daily news and social media (as I am prone to do), I have to remind myself that I am building a skewed model. Embracing the long view has helped me step outside these habitual behaviours, allowing me to think and see more clearly. To be clear, this is not a justification to turn away from trusted sources, but a reason to supplement them with what I call *long news*: long-term trends or lesser-known histories, as well as speculative futures from experts that go beyond the typical techno-consumerist visions. Through long news comes greater insight, and fewer feelings of fear, outrage and helplessness.

THE LONG VIEW PROVIDES A CLEARER PICTURE OF PROGRESS

Our most privileged ancestors would have marvelled at the existence that the average person on Earth now enjoys. We have a quality of life that Cleopatra, William the Conqueror or Louis XVI could only have dreamed of. Theirs was a time of greater violence, prejudice and illness. Imagine what they would have made of flushing toilets, electric fridges, the internet, high-yield wheat, human rights law, universal suffrage, free schooling for children, and vaccines.

However, our scientific and technological achievements do not mean we have reached the peak of human potential or enlightenment. Nor are all things universally better. Indeed, many discoveries and technologies have made the world worse: promoting inequalities, intensifying warfare, harming nature, fostering hate or hastening self-destruction.[2] If you take the perspectives of certain groups of oppressed peoples, progress has brought collapse and enslavement; and for many non-human animals, it has led to total catastrophe.

The long view offers a lens to see both the improvements and the mistakes, and therefore a more nuanced sense of what progress can – and should – mean. We needn't cling to single narratives that the world is unequivocally getting better or getting worse: it is both, depending on what metric you are looking at and the vantage point you adopt. Doing so provides a base of evidence on which to act, make reparations and move forward.

THE LONG VIEW IS AN ENGINE FOR HOPE

When I talk or write about the long term – particularly the deep future – I sometimes encounter a form of resigned nihilism, expressed as: 'That's all very well, but we'll all be gone.' In some circles, it has become almost fashionable to quip that the apocalypse is nigh. In that context, it becomes something of a trap to talk about the long view: you're either seen as far too optimistic or callously unconcerned with today's problems. Too much pessimism, though, can lead to doomism – a perspective that remains locked in the present, mired in apathy, helplessness or anger.

It's undeniable that there are severe dangers ahead for the world. However, I would prefer to navigate those futures equipped with the belief that I and those around me have agency. I do not believe I can steer a path alone – it is a collective, cross-generational enterprise – nor do I know how tomorrow's events will play out. But I would rather operate under the belief that I can do something rather than nothing. Surprisingly, most of the scholars I have met who actually study the end of the world are not doomist at all. They might spend their days contemplating catastrophe, but they believe that a long-term perspective is necessary to avoid that fate, and are often motivated by the prospect of future flourishing.

The blinkered norms of our age hide so much from us. What we mustn't do is let them hide the possibility that things can improve as well as worsen. When past generations faced gross inequalities, the fog of conflict, or great injustices, it would have felt overwhelming at times – but over generational timescales, changes for the better are not impossible, and in this we can find a source of energy and determination.

There's a term used by some long-term thinkers that I believe deserves to be known more widely: *existential hope*. This is the opposite of exist-ential catastrophe: it's the idea that there could be radical turns for the better, so long as we commit to bringing them to reality. Existential hope is not about escapism, utopias or pipe dreams, but about *preparing the ground*: making sure that opportunities for a better world don't pass us by.

So, if taking the long view demands anything of us, it is this: a com-mitment to seeking and cultivating hope when all feels bleak. This may well prove to be the grandest challenge of our time, but it is what we owe to our predecessors and our descendants.

Over the long term, I believe it is within humanity's potential to build towards what I call a *deep civilisation*. If a time-blinkered society is one that cannot escape the present moment, a deep civilisation is one that has a richer sense of its roots, and its trajectory into tomorrow.

In a deep civilisation, businesses are not swayed by short-term individualist profits, but are motivated by ethical and sustainable goals. Politicians have the foresight and wisdom to support policies that benefit all people and living creatures in all times, not just their own voter base. Journalists and communicators provide temporal context and depth, rather than outrage and noise. Technologists and designers aim to foster cross-generational connection, not anger and division. And every citizen knows that they are each a link in a chain that stretches across the generations, with the collective capability to improve the world for their children. They are, to quote the scientist Jonas Salk, 'good ancestors'.[3]

At the same time, each member of this deep civilisation is keenly aware that their evolution is incomplete – that the societies and communities they are building are just one step on the way towards what they have the potential to become. They pass that hope across generations with the promise of a world that could be more just, wiser and more enlightened.

Across the very long term, there could be whole new heights for our species that we have yet to explore. As the philosopher Toby Ord has argued, our present may be marred by suffering and unsustainable practices but we should aspire to more than ridding these problems. 'A world without agony and injustice is just a lower bound on how good life could be. Neither the sciences nor humanities have yet found any upper bound,' he writes. 'We get some hint at what is possible during life's best moments: glimpses of raw joy, luminous beauty, soaring love. Moments when we are truly awake. These moments, however brief, point to possible depths of flourishing far beyond the status quo, and far beyond our current comprehension.'[4]

Such talk of a grander future may sound utopian and out of reach – but it is not impossible, so long as we make the right choices today. One reason it is so difficult to see how things could be so different is a psychological effect called the 'end of history' illusion, which describes how people struggle to imagine how they might change later in their lives.[5]

While people accept that they have changed significantly since they were children, they assume that their present self is how they will always be. This can be a collective illusion too, hiding from us how much our societies still stand to evolve.[6]

Are we ready to become a deep civilisation? Not yet. There is still a long way to go before we fully understand and accept the relationships between the human lifespan, past and future generations, and the long-term timescales of Earth and the natural world. And for the foreseeable future, deep time may continue to feel daunting and intractable for many: a 'delightful horror' that feels too big to imagine with our current mental faculties. Humanity, though, is nothing if not flexible: throughout our evolution we have adapted and expanded our perspective to embrace all sorts of abstract and complex concepts that exist outside direct experience and memory: morality, peace, charity, freedom and law – to name but a few. Across history, we have encountered ever-more complex ideas, and learned ways to break them down into terms and concepts that we *could* understand.

Crucially, when we seek the long view, we do not do so alone. As a social species, we build on the minds and experiences of others – past and present. And through this cooperation, an individual can unlock insights that they cannot see, hear or feel themselves.

So, the coming years could mark a turning point in our temporal evolution. On one path, we destroy our species because we failed to think long term; on the other, we flourish into a future extending millions of years hence. If we are to thrive beyond the next century, we must transform our relationship with time – to close the gap between the salient experience of the present moment and the far brighter trajectories that could lie ahead.

As I neared the end of writing this book, I asked my daughter Grace once again about time – specifically, whether she feels pessimistic or optimistic about the future. Reassuringly, she answered the latter. Now she is nine, she is increasingly aware of twenty-first-century problems, such as climate change and social injustice. However, I often draw hope from her sense of agency, possibility and fun.

'When I started my book, I imagined you when you'll be eighty-six years old, living in the twenty-second century,' I said. 'How do you think it will it feel to be that age, looking back? What do you think you'll be doing?'

She paused the game on her screen, and thought about it for a moment.

'I imagine me and my friend Sarah, living in a retirement home in space, sitting on rocking chairs,' she said, before adopting a voice like an old lady. 'We're saying "back in my day . . . phones weren't holograms".'

I still don't know what to make of this, but it made me smile.

Acknowledgements

When I was giving the groom's speech at my wedding, back in 2008, I relied heavily on cards to remind me of the many people I wanted to thank – our extended families, the bridesmaids, the best men and more. When I got to the end of the cards, which by then were slightly damp from my nerves, I indicated that I was about to wrap up the speech. There was an intake of breath in the room: I realised I was about to sit down without saying anything about the most important person: my wife, Kristina. The card I'd carefully written about her had somehow been lost. So, I ad-libbed and spoke from the heart. Turns out I didn't need notes to say how I felt.

With that experience in mind, I owe it to Kristina to start my thanks with her, along with my daughter, Grace. As I wrote in the book's opening pages, this project began with them: when we started our family, it altered how I thought about generational time. But I owe them so much more: this book wouldn't exist without their encouragement, inspiration, patience and love. Years ago, it was Kristina who nudged me to take my jumbled thoughts about the long view and reflect more deeply. When I was uncertain about whether to embark on a multi-year project on a single topic, she asked me a simple question: *'What's always been there?'* This query stuck in my mind, and encouraged me to join the dots between experiences and knowledge that spanned my life. It also gave me the confidence to know that the long view had, and always will be, a lifelong interest. Over the years since, Kristina has listened patiently and insightfully to me talking about the book, for hours on end, and has undoubtedly made it better. (For my forty-first birthday, she gifted me an annotated binder of detailed, inspiring feedback, after secretly reading the draft cover to cover.) Along the way, there have been difficult years for our family, facing illness and the loss of our son, Jonah, but I am reminded every day of how lucky I am to know my wife and daughter: they are the lights in my life, and I love them very much.

The Long View has been well over five years in the making, and along the way, probably hundreds of people have helped to shape its final form. Attempting to be comprehensive in my thanks is therefore a fraught endeavour – especially from someone who almost forgot to talk about his wife at his wedding – but here goes . . .

First, thank you to Kate Evans at Peters, Fraser and Dunlop. After reading my work at the BBC, she was the first person to independently recognise the potential for a book about the long view, and took a leap of faith with me when the idea was just a few pages of notes. She has been an inspiring supporter of the book ever since. Her unguarded Australian enthusiasm perfectly balances my more sober, serious British tendencies, and I'm really glad she's on my team.

I also am indebted to my editor Lindsay Davies; Alex Clarke, the publishing director of the Wildfire imprint; and everyone else who has supported the book at Headline. Both Lindsay and Alex understood and backed my vision from the very start, helping me define a positive and hopeful framing, and lending clarity and direction to my writing at every stage. They identified gaps I had not considered, thickets I had not untangled, and statements that ought to be amplified. One of the more memorably concise suggestions? *'Be more shouty here.'* They were, as always, correct.

Another person to whom I owe a great deal of thanks is Toby Tremlett, who worked with me as a research assistant over many months –

researching, fact-checking, drafting and much more. His findings, ideas, advice and writing brought far more depth and rigour than I could have managed on my own, and our regular catch-ups motivated me to keep going during the mid-project doldrums when I felt far from land. Toby's time was funded by an unexpected grant from Open Philanthropy, which was proposed by Claire Zabel – I am extremely grateful for this support. (It also wouldn't have happened without an introduction by Michelle Hutchinson at 80,000 Hours, so thank you to her too.)

The foundational research for the book came from an opportunity I still feel incredibly lucky to have experienced. Between 2019 and 2020, I took a sabbatical from the BBC to spend an academic year at MIT on the Knight Science Journalism fellowship, where I was given rare space and time to develop the book's scope. Deborah Blum (author of *The Poison Squad* and more), Ashley Smart, Bettina Urcuioli and the rest of the KSJ team made me and my family feel truly welcome throughout our time in Cambridge, so thank you deeply to all of them – it was one of the best years of my life. Thank you also to the wonderful 2019–20 KSJ fellows: Andrada Fiscutean, Anil Ananthaswamy (*Through Two Doors at Once* and others), Bethany Brookshire (*Pests*), Eva Wolfangel (*Ein Falscher Klick*), Jon Fauber, Molly Segal, Sonali Prasad, Thiago Medaglia and Tony Leys. During this time, I also learned a lot from former KSJ fellow Rachel Gross, who was roughly a year ahead of me writing her book, *Vagina Obscura*. Finally, thank you to the academics at MIT and Harvard who gave me their time and allowed me to audit their classes: in particular Kieran Setiya (*Life is Hard*) and Joshua Greene (*Moral Tribes*), who gave me a rapid introduction to philosophy and ethics, Giovanni Bazzana on apocalypses, and Rebecca Saxe, who teaches an outstanding seminar on the neuroscience of morality.

From proposal to manuscript, various readers have generously given their time and feedback, including the KSJ fellows mentioned above (Andrada, Anil, Bethany, Rachel, Sonali), Deborah Blum, Michael Haydock, Chris Daniel and Lauren Holt. David Robson's (*The Intelligence Trap*, *The Expectation Effect*) suggestions provided tonnes of helpful new material I would otherwise I have missed, and he has been a good friend and advisor from the very start. Thomas Moynihan (*X-Risk*) generously read the manuscript cover to cover, providing reams of notes, constructive criticisms, fact-checks and motivating encouragement. As did Luke Kemp – aka 'positive feedback Luke' – whose comments were so thorough and on point that he successfully managed to get me to cut an entire chapter that shouldn't have been there. Luke, along with Lauren, have also been much-valued writing partners in our Friday 'Calliope Club' as they work on their own long-term projects – I think we probably do more talking than writing, but it has been a great source of weekly motivation. I also wish to express a major thank-you to a colleague of theirs at CSER in Cambridge: Matthijs Maas. Over many months, Matthijs sent me dozens of papers, articles and ideas that fed into the book, with no expectation other than he enjoys sharing knowledge. The same goes for Stefan Schubert, who was similarly generous with his reading recommendations, and Anders Sandberg, who has a truly infectious sense of curiosity and fun.

I work with a wonderful team at the BBC, who have been a continual source of inspiration and camaraderie over the years. There isn't space to name the whole team and all our terrific freelancers, but for making the book happen, I am particularly grateful for the support of Richard Gray, Amanda Ruggeri, Simon Frantz and Mary Wilkinson – before, during and after my career break at MIT. And Jon Fildes, now at DeepMind, who has been a champion of my work for a decade.

Over the years, a number of book authors not already mentioned have kindly shared their experience, knowledge and war stories about publishing, something that, as a first-time author, is so very valuable. They include: Tom Chatfield (*How to Think* and others), Vincent Ialenti (*Deep Time Reckoning*), Roman Krznaric (*The Good Ancestor* and others), David Farrier (*Footprints*), Bina Venkataraman (*The Optimist's Telescope*), Melissa Hogenboom (*The Motherhood Complex*), Rutger Bregman (*Humankind, Utopia for Realists*), Toby Ord (*The Precipice*), Will MacAskill (*Doing Good Better, What We Owe the Future*), Caspar Henderson (*A New Map of Wonders* and others), Simon Parkin (*The Island of Extraordinary Captives*), Rachel

Nuwer (*Poached*), Alok Jha (*The Water Book*), Helen Thomson (*Unthinkable*), Sam Arbesman (*Overcomplicated*), Rowan Hooper (*How to Spend a Trillion Dollars*) and Sally Adee (*We Are Electric*).

There are many others who have shaped my thinking and ideas over the years, directly and indirectly. For a full list of sources, see the endnotes, but in no particular order, here is a selection of other notable influences and fellow 'long viewers' who have inspired me with their work or advice: Beatrice Pembroke, Ella Saltmarshe, Nicholas Paul Brysiewicz, Alexander Rose, Stewart Brand, Ahmed Kabil, George Gantz, Simon Caney, Kathy Peach, Michael Ogden, Peter Dean, Simon Bray, Richard Sandford, Philipa Duthie, Cristina Parreño, Rina Tsubaki, Zaria Gorvett, James Janson Young, Marcia Bjornerud, Katie Paterson, Rebecca Altman, Andy Russell, Lee Vinsel, Cat Tully, Sophie Howe, Lord Martin Rees, John Boyd, Lucien Holscher, Oliver Burkeman, Seth Baum, Nick Bostrom, Natalie Cargill, Tyler John, Fin Moorhouse, Arden Koehler and Garrison Lovely. Thank you also to Dave Price, Dominic Jarvis, Hannah Davies and João Duarte for allowing me to use their photos and graphics, and to Nigel Hawtin, who skilfully helped me visualise the scale of future generations back in 2019.

Finally, thank you to my parents, as well as the wider Fisher and Jackets families. We owe a great deal to the lives of those who came before us, but words cannot express how grateful I am to my mum, Jennifer, and my dad, Clive, for giving me the freedom to choose how I live mine.

Wildfire Books would like to thank everyone at Headline who helped to publish *The Long View* in the UK.

Editorial
Alex Clarke
Lindsay Davies
Areen Ali

Picture research
Cathie Arrington

Contracts
Helen Windrath

Sales
Rebecca Bader
Chris Keith-Wright
Rachel Campbell

Rights
Rebecca Folland
Grace McCrum
Ruth Case-Green
Lakhesia Adams Poku

Design
Patrick Insole

Production
Louise Rothwell

Publicity
Joe Thomas

Marketing
Joe Yule

Copy Editor
Vicky Orchard

Proofreader
Sarah Coward

Indexer
Alan Rutter

Picture Credits

Page 26: Hutton's Unconformity at Inchbonny, Jedburgh. Illustration by John Clerk of Eldin from James Hutton, *Theory of the Earth*, in *Transactions of the Royal Society of Edinburgh*, Volume I, 1788. (Natural History Museum/Alamy)

Page 121: *Landscape with the Fall of Icarus*, oil on canvas by or after Pieter Bruegel the Elder, *c*.1560. Royal Museums of Fine Arts of Belgium (Artefact/Alamy)

Page 137: John Mainstone/Pitch Drop Experiment (© The University of Queensland)

Page 184: Tree branches (© Richard Fisher)

Page 186: Rocks on the wall (© Richard Fisher)

Page 197: The previous and new Naiku complex of Ise Grand Shrine (Kyodo News/Newscom/Alamy)

Page 198: *The Official Move to the Rebuilt Ise Shrine*. Woodblock print by Utagawa Kuniyoshi, 1849. (Photograph © Museum of Fine Arts, Boston. All rights reserved. / Bridgeman Images)

Page 225: The diagram showing the scale of the unborn is based on a graphic produced by Nigel Hawtin for BBC Future

Page 242: Stripes cross-section (© Richard Fisher)

Page 242: Supercontinent Novopangea (adapted from H. Davies et al.)

Page 274: White Horse of Uffington (© Dave Price)

Notes

Introduction: A Longer View

1   Burke, Edmund, *Reflections on the Revolution in France* (J. Dodsley, 1790).

2   'The Deep Future: A Guide to Humanity's Next 100,000 Years', *New Scientist* (2012).

3   'Deep Civilisation' series, BBC Future (2019).

4   Note that 'longtermism', written without a hyphen, is not the same as 'long-termism'. The former has a specific definition, described in Part III.

1. A Brief History of Long Time

1   Gellner, Ernest, *Thought and Change* (Weidenfeld & Nicolson, 1964).

2   Lloyd, G., 'Foresight in Ancient Civilisations', in Sherman, Lawrence W., and Feller, David Allan (ed.), *Foresight* (Cambridge University Press, 2016).

3   Gellner (1964).

4   Damon, C., 'Greek Parasites and Roman Patronage', *Harvard Studies in Classical Philology* (Harvard University Press, 1995).

5   Extract from Gatty, Alfred and Margaret, *The Book of Sun-dials* (George Bell & Sons, 1900). The extract from Plautus originally appeared in the book of the Roman writer Aulus Gellius, *Noctes Atticae* (*Attic Nights*), Book 3, Chapter 3.

6   Foster, R., 'Biological Clocks: Who in This Place Set Up a Sundial?', *Current Biology* (2012).

7   Shaw, B., 'Did the Romans have a future?', *The Journal of Roman Studies* (2019).

8   Moynihan, Thomas, *X-Risk: How Humanity Discovered Its Own Extinction* (MIT Press, 2020).

9   Lloyd, G., 'Foresight in Ancient Civilisations', Darwin College Lecture Series.

10   The Prediction Project (2020) Roman Augury.

11   Tacitus, *The Annals: The Reigns of Tiberius, Claudius, and Nero*, translated by Yardley, J. C. (Oxford University Press, 2008).

12   A few have challenged this interpretation, such as Shushma Malik, *The Nero-Antichrist* (Cambridge University Press, 2020).

13   Dickinson, Emily, 'Forever – is composed of Nows' (690), in Franklin, R. W. (ed.), *The Poems of Emily Dickinson* (Harvard University Press, 2005).

14   According to Giovanni Bazzana and colleagues at Harvard University, who teach a course on the apocalypse, which I audited in 2019.

15   Villarreal, Alexandra, 'Meet the doomers: why some young US voters have given up hope on climate', *Guardian* (2020).

16   Waldron, A., 'The Problem of the Great Wall of China', *Harvard Journal of Asiatic Studies* (1983).

17   Corrigan, I., *Stone on Stone: The Men Who Built the Cathedrals* (The Crowood Press, 2018).

18   Some sources attribute this prayer to Lincoln, others to Winchester: see Miller, Kevin, 'God's glory in wood and stone', *Christian History* (1996); Corrigan (2018).

19   As well as Hölscher (see next note), I owe this insight to Lord Martin Rees.

20   A good deal of the history in this chapter is built on the writing of the historian Lucian Hölscher of Ruhr-University Bochum. See: Hölscher, L., 'Future Thinking: A Historical Perspective', in Oettingen, Gabrielle, Timur, Sevincer, and Gollwitzer, Peter (eds.), *Psychology of Thinking About the Future* (Guilford Press, 2019).

21   Various writings of the historian Reinhart Koselleck, such as 'Social History and Conceptual History', *International Journal of Politics, Culture and Society* (1989).

22   Burke, P., 'Foreword', in Brady, A., and Butterworth, E., *The Uses of the Future in Early Modern Europe* (Routledge, 2010).

23 Lutz, W., Butz, W., and Samir, K. C., *World Population & Human Capital in the Twenty-First Century: An Overview* (Oxford University Press, 2014); and King, G., *Natural and Political Observations and Conclusions Upon the State and Condition of England* (1696).

24 UN 2019 World Population Prospects.

25 Johnston, Warren, *Revelation Restored: The Apocalypse in Later Seventeenth-century England* (Boydell Press, 2011).

26 Snobelen, S., '"A time and times and the dividing of time": Isaac Newton, the Apocalypse, and 2060 A.D.', *Canadian Journal of History* (2016).

27 'Siccar Point', The Geological Society (accessed February 2020).

28 Cuvier, Georges, *Essay on the Theory of the Earth* (Kirk & Mercein, 1813).

29 Hutton, J., 'Theory of the Earth', *Transactions of the Royal Society of Edinburgh* (1788).

30 Hölscher (2019).

31 Kant, Immanuel, *Allgemeine Naturgeschichte und Theorie des Himmels* (1755), translated by Johnston, Ian, *Universal Natural History and Theory of the Heavens* (Richer Resource Publications, 2008).

32 Kant, Immanuel, *Beantwortung der Frage: Was ist Aufklärung?* (1784), translated by Nisbet, H. B., *An Answer to the Question: 'What is Enlightenment?'* (Penguin, 2013).

33 Alkon, Paul, *Origins of Futuristic Fiction* (UGA Press, 1987).

34 Alkon, P., 'Samuel Madden's "Memoirs of the Twentieth Century"', *Science Fiction Studies* (1985).

35 *The Book Challenged: Heresy, Sedition, Obscenity* (2009). Exhibition at the University of Otago, New Zealand.

36 Moynihan (2020).

37 Carlyle, Thomas, 'Boswell's *Life of Johnson*' (1832).

38 Campbell, Thomas, *Life and Letters of Thomas Campbell* (Hall, Virtue & Company, 1850).

39 Mumford, Lewis, *Technics and Civilization* (University of Chicago Press, 1934).

40 Adam, Barbara, *Timescapes of Modernity: The Environment and Invisible Hazards* (Routledge, 1998).

41 Ivell, D., 'Phosphate Fertilizer Production – From the 1830's to 2011 and Beyond', *Proceedia Engineering* (2012).

42 Hölscher (2019).

43 H. G. Wells's talk's title was *The Discovery of the Future*, which Lucian

Hölscher is alluding to in his description of the eighteenth-century long view.

44   Moynihan, Thomas, 'Creatures of the dawn: How radioactivity unlocked deep time', BBC Future (2021).

45   Wells, H. G., *A Short History of the World* (Cassell & Company, 1922).

46   Orwell, George, 'Wells, Hitler and the World State', *Horizon* (1941).

47   Guse, J., 'Volksgemeinschaft Engineers: The Nazi Voyages of Technology', *Central European History* (2011).

48   Deutsch Gabel, now Jablonné v Podještědí in the Czech Republic.

49   Meier, C., 'Consigning the Twentieth Century to History: Alternative Narratives for the Modern Era', *The American Historical Review* (2000).

50   Novak, Matt, '42 Visions for Tomorrow from the Golden Age of Futurism', *Gizmodo* (2015).

51   As described by Aleida Assmann (see next note).

52   Assmann, Aleida, 'Transformations of the Modern Time Regime', in Lorenz, C., and Bevernage, B., *Breaking up Time: Negotiating the Borders Between Present, Past and Future* (Vandenhoeck & Ruprecht, 2013).

53   Assmann, Aleida, *Is Time out of Joint?: On the Rise and Fall of the Modern Time Regime* (Cornell University Press, 2020).

54   Brown, Kimberly, *The I-35W Bridge Collapse: A Survivor's Account of America's Crumbling Infrastructure* (Potomac Books, 2018).

55   Fisher, T., 'Fracture-Critical: The I-35W Bridge Collapse as Metaphor and Omen' in Nunnally, Patrick (ed.), *The City, the River, the Bridge: Before and After the Minneapolis Bridge Collapse* (University of Minnesota Press, 2011).

56   Jordheim, H., and Wigen, E., 'Conceptual Synchronisation: From Progress to Crisis', *Millennium: Journal of International Studies* (2018).

57   Hartog, François, *Régimes d'historicité: présentisme et expériences du temps* (Seuil, 2003), translated by Brown, Saskia, *Regimes of Historicity: Presentism and Experiences of Time* (Columbia University Press, 2016).

58   Hartog's sociological definition is different to that of 1. historical presentism, which describes looking at historical events through the lens of present-day norms, and 2. philosophical presentism, which essentially proposes that only present things exist.

59   Gumbrecht, Hans Ulrich, *Our Broad Present: Time and Contemporary Culture* (Columbia University Press, 2014).

60   Tamm, M., 'How to reinvent the future?', *History and Theory* (2020); Esposito, Fernando, ed., *Zeitenwandel: Transformationen geschichtlicher*

Zeitlichkeit nach dem Boom (Vandenhoeck and Ruprecht, 2017); Tamm, Marek, and Olivier, Laurent, ed., *Rethinking Historical Time: New Approaches to Presentism* (Bloomsbury Academic, 2019).

61   Baschet, Jérôme, *Défaire la tyrannie du present: Temporalités émergentes et futurs inédits* (La Découverte, 2018).

62   Gilbert, Daniel, *Stumbling on Happiness* (Alfred A. Knopf, 2006).

63   Hartog (2003), trans. Brown (2016).

2. Selling Short

1   Rae, John, *The Sociological Theory of Capital: Being a Complete Reprint of the New Principles of Political Economy*, 1834 (Macmillan, 1905).

2   Keynes, John Maynard, *The General Theory of Employment, Interest and Money* (Macmillan & Co., 1936).

3   Favre, D., 'The Development of Anti-Cruelty Laws During the 1800's', *Detroit College of Law Review* (1993).

4   Lubinski, Christina, 'Fighting Friction: Henry Timken and the Tapered Roller Bearing', *Immigrant Entrepreneurship* (2011).

5   Hobbs Pruitt, Bettye, *Timken: From Missouri to Mars – a Century of Leadership in Manufacturing* (Harvard Business Press, 1998).

6   Schwartz, Nelson D., 'How Wall Street Bent Steel', *New York Times* (2014). I owe many of the details of the Timken story to Schwartz's excellent reporting in this piece.

7   'Timken steel spinoff proposal still on table following meeting', *Akron Beacon Journal* (2014).

8   Benoit, D., et al, 'Relational Investors Plans to Wind Down Operations, Dissolve Current Funds', *Wall Street Journal* (2014).

9   Pritchard, Edd, 'Timken Steel job cuts continue as cost-cutting measures expand', *Canton Repository* (2019).

10   Timken company website (accessed September 2020).

11   Fortado, Lindsay, 'Companies faced more activist investors than ever in 2019', *Financial Times* (2019).

12   Maloney, T., and Almeida, R., *Lengthening the Investment Time Horizon*, MFS White Paper (2019).

13   *Corporate Longevity: Index Turnover and Corporate Performance*, Credit Suisse (2017).

14   *2021 Corporate Longevity Forecast*, Innosight (2021).

15   de Geus, Ari, *The Living Company* (Nicholas Brealey, 1999).

16   'Corporate Long-term Behaviors: How CEOs and Boards Drive Sustained Value Creation', FCLT Global (2020).

17   'Predicting Long-Term Success for Corporations and Investors Worldwide', FCLT Global (2019).

18   Goodwin, Crauford, *Maynard and Virginia: A Personal and Professional Friendship* (History of Political Economy, 2007).

19   Osterhammel, Jurgen, *The Transformation of the World: A Global History of the Nineteenth Century* (Princeton University Press, 2009).

20   Keynes, John Maynard, *The General Theory of Employment, Interest and Money* (Macmillan & Co., 1936).

21   Wasik, John F., *Keynes's Way to Wealth: Timeless Investment Lessons from the Great Economist* (McGraw-Hill Education, 2013).

22   Kraft, A., et al., 'Frequent Financial Reporting and Managerial Myopia', *The Accounting Review* (2018).

23   'Short-termism Revisited', CFA Institute (2021).

24   'Considerations on COM (2011)683', EU Monitor (2013).

25   Unilever Sustainable Living Plan (2010).

26   Skapinker, Michael, 'Corporate plans may be lost in translation', *Financial Times* (2010).

27   Ignatius, Adi, 'Captain Planet', *Harvard Business Review* (2012).

28   *Going Long Podcast: Paul Polman*, FCLT Global (2020).

29   Graham, J. R., et al., 'Value Destruction and Financial Reporting Decisions', *Financial Analysts Journal* (2006). Another study by McKinsey and Company and FCLT Global asked a similar question, and got a response of 60 per cent who said they have cut discretionary spending or delayed projects, among others to meet quarterly promises (Barton, B., and Zoffer, J., 'Rising to the Challenge of Short-termism', FCLT Global (2016))

30   Martin, Roger L., 'Yes, short-termism really is a problem', *Harvard Business Review* (2015).

31   Mauboussin, M. J., and Callahan, D., 'A Long Look at Short-Termism: Questioning the Premise', Credit Suisse (2014).

32   Murray, Sarah, 'How to take the long-term view in a short-term world', *Financial Times* (2021).

33   'Three Girls Gone: The Ford Pinto and Indiana v. Ford Motor Co', *Orangebean Indiana* (2019).

34 There's nuance to the Pinto story that I have abridged, and a few myths too. See: Vinsel, L., 'The Myth of the "Pinto memo" is Not a Hopeful Story for Our Time', *Medium* (2021); Lee, M., and Ermann, M. D., 'Pinto "Madness" as a Flawed Landmark Narrative: An Organizational and Network Analysis', *Social Problems* (1999).

35 Opening Statement of Senator Carl Levin, US Senate Permanent Subcommittee on Investigations, *Wall Street and the Financial Crisis: The Role of Credit Rating Agencies* (2010).

36 Mauboussin and Callahan (2014).

37 'Predicting Long-Term Success for Corporations and Investors Worldwide', FCLT Global (2019).

38 Kamga, C., Yazic, M. A., and Singhal, A., 'Hailing in the Rain: Temporal and Weather-Related Variations in Taxi Ridership and Taxi Demand-Supply Equilibrium', Transportation Research Board 92nd Annual Meeting (2013).

39 Henry, J. F., 'A Neoliberal Keynes?', *International Journal of Political Economy* (2018).

40 Mazzucato, Mariana, and Jacobs, Michael, *Rethinking Capitalism: Economics and Policy for Sustainable and Inclusive Growth* (Wiley, 2016).

41 Barton, Dominic, 'Capitalism for the Long Term', *Harvard Business Review* (2011).

42 Skapinker, Michael, 'Unilever's Paul Polman was a standout CEO of the past decade', *Financial Times* (2018).

43 Barton, Dominic et al., 'Measuring the Economic Impact of Short-termism', McKinsey Global Institute (2017).

44 Bushee, Brian, 'Identifying and Attracting the "Right" Investors: Evidence on the Behavior of Institutional Investors', *Journal of Applied Corporate Finance* (2005).

45 Maboussin, Michael J., and Rappaport, Albert, 'Reclaiming the Idea of Shareholder Value', *Harvard Business Review* (2016).

46 Brochet, F., Serafeim, G., and Loumioti, M., 'Speaking of the Short-Term: Disclosure Horizon and Managerial Myopia', *Review of Accounting Studies* (2015).

47 Son, Masayoshi, *Softbank Next 30-year Vision* (2010).

48 Nationwide 'long-established company' survey, Tokyo Shoko Research (2016)

49 O'Halloran, Kerry, 'The Adoption Process in Japan' in *The Politics of Adoption* (Springer, 2015).

50   Mehrota, V., et al., 'Adoptive Expectations: Rising Sons in Japanese Family Firms', *Journal of Financial Economics*, (2013).

51   'The Long-term Habits of Highly Effective Corporate Boards', FCLT Global (2019).

52   de Geus (1999).

53   Rose, Alexander, 'The Data of Long-lived Institutions', The Long Now Foundation (2020).

54   Sasaki, Innan, 'How to build a business that lasts more than 200 years – lessons from Japan's shinise companies', *The Conversation* (2019).

55   O'Hara, W. T., *Centuries of Success: Lessons from the World's Most Enduring Family Businesses* (Avon: Adams Media, 2004).

56   It is now a subsidiary of Takamatsu Construction, after it was acquired in 2006.

57   Taleb, Nassim Nicholas, *Antifragile: Things That Gain from Disorder* (Penguin, 2012).

3. Political Pressures

1   Hamilton, Alexander, *The Federalist Papers: No. 71* (1788).

2   de Tocqueville, Alexis, *Democracy in America* (Saunders and Otley, 1838).

3   Greider, William, 'The Education of David Stockman', *The Atlantic* (1981).

4   Johnson, Haynes, 'Stockman's Economy an Intricate Puzzle, Without Any People', *Washington Post* (1981).

5   Greider (1981); I owe the discovery of this notorious comment (made during an interview with journalist William Greider) to the political scientist Simon Caney – see next note.

6   Caney, S., *Democratic Reform, Intergenerational Justice and the Challenges of the Long-Term*, CUSP essay series on the Morality of Sustainable Prosperity (2019); Caney, S., 'Political Institutions for the Future: A Fivefold Package', in González Ricoy, Iñigo, and Gosseries, Axel (eds.), *Institutions for Future Generations* (Oxford University Press, 2016); Pierson, P., *Politics in Time: History, Institutions, and Social Analysis* (Princeton University Press, 2004).

7   Stockman, David A., *The Triumph of Politics: Why the Reagan Revolution Failed* (Harper & Row, 1986).

8   Caney (2019); Friedman, Thomas L., 'Obama on Obama on Climate', *New York Times* (2014); 'The Quest for Prosperity', *The Economist* (2007).

9   Lempert, R., 'Shaping the Next One Hundred Years: New Methods for Quantitative, Long-Term Policy Analysis', *Rand* (2007).

10 E.g. Norway's Government Pension Fund Global for the long-term management of revenue from its oil and gas reserves. See: www.nbim.no/en

11 Luna, Taryn, 'Winter storms impose high costs for business', *Boston Globe* (2015).

12 Dudley, David, 'Snowstorm Mayors: Don't Blow This', *Bloomberg Citylab* (2017).

13 Caney (2019).

14 The framing of political problems as 'uncinematic' comes from Rob Nixon's concept of 'slow violence': Nixon, Rob, *Slow Violence and the Environmentalism of the Poor* (Harvard University Press, 2012).

15 You could also call them 'wildfires' but given that wildfires themselves are a threat of climate change, I elected not to.

16 Cohen, J., et al., 'Divergent consensuses on Arctic amplification influence on midlatitude severe winter weather', *Nature* (2020). A caveat: cause/effect is still to be fully untangled. The warming in the Arctic is known, but what its effects will be elsewhere is difficult to predict with certainty.

17 'What climate change means for Massachusetts', Environmental Protection Agency (2016).

18 Massachusetts Energy and Environment Performance Review & Recommendations for Governor Baker's Second Term (2019).

19 I owe the conceptual inspiration for this framework to Stewart Brand's 'pace layers', which depict the different rates of change within a society. See: Brand, S., 'Pace Layering: How Complex Systems Learn and Keep Learning', *Journal of Design and Science* (2018).

20 Jordheim, H., and Wigen, E., 'Conceptual Synchronisation: From Progress to Crisis', *Millennium: Journal of International Studies* (2018).

21 It's not that politicians in history did not face slow-paced problems, but my argument is that human progress and technological complexity has created many more than would occur otherwise. And when they did exist, such as pollution in the Industrial Revolution or intergenerational poverty, then there was less awareness of long-term consequences.

22 The average term is 4–5 years, but a handful of democracies have longer terms of 7 years, such as Ireland's Taoiseach, Italy's Prime Minister and Israel's President.

23 Offe Claus, *Europe Entrapped* (Polity, 2015).

24 Caney (2019).

25 Thanks to Luke Kemp of Cambridge University for this example.

26 No elections, but since 1982 the Chinese constitution has given leaders fixed terms. There's been recent speculation that Xi Jinping may seek to extend his time in office, however.

27 Scobell, A., et al, 'China's Grand Strategy: Trends, Trajectories, and Long-Term Competition', *RAND Research Report* (2020).

28 McQuilken, J., 'Doing Justice to the Future: A global index of intergenerational solidarity derived from national statistics', *Intergenerational Justice Review* (2018).

29 Krznaric, Roman, *The Good Ancestor* (W. H. Allen, 2020); Intergenerational Solidarity Index, romankrznaric.com (2020).

30 Chen, A., Oster, E., and Williams, H., 'Why Is Infant Mortality Higher in the United States Than in Europe?', *American Economic Journal: Economic Policy* (2016).

31 Dijkstra, Erik, 'The strengths of the academic enterprise', in Broy, M., and Schieder, B. (ed.), *Mathematical Methods in Program Development* (Springer, 1997).

32 Two of the most commonly cited dates in forecast reports are 2050 and 2100.

33 Aizenberg, E., and Hanegraaff, M., 'Is politics under increasing corporate sway? A longitudinal study on the drivers of corporate access', *West European Politics* (2019).

34 'Ezra Klein on aligning journalism, politics, and what matters most', *80,000 Hours* podcast (2021).

35 Rusbridger, Alan, 'Climate change: why the Guardian is putting threat to Earth front and centre', *Guardian* (2015).

36 Klite, P., Bardwell, R., and Salzman, J., 'Local TV News: Getting away with Murder', *The International Journal of Press and Politics* (1997).

37 Johnson, Boris, 'This cap on bankers' bonuses is like a dead cat – pure distraction', *Daily Telegraph* (2013).

38 Jefferson, T., 'Letter to John Taylor' (1816); Mill, J. S., *Hansard* (Volume 182, 1866); Marx, Karl, *Das Kapital: Kritik der politischen Ökonomie, Buch III* (Otto Meisner, 1894), translated by Fernbach, David, *Capital: A Critique of Political Economy, Volume Three* (Penguin, 1992).

39 Englander, John, 'Applying Jacque Cousteau's wisdom', *Think Progress* (2010).

40 'One Man's Mission: Pierre Chastan', The Cousteau Society (2001); *Meeting*

of Secretary-general with Cousteau Society to Receive Petition on 'Rights of Future Generations', United Nations (2001).

41   *Declaration on the Responsibilities of the Present Generations Towards Future Generations; Draft Declaration on the safeguarding of future generations*, United Nations (1997).

42   'Intergenerational Solidarity and the Needs of Future Generations', UN Report of the Secretary General (2013).

43   'Our Common Agenda', United Nations (2021).

44   Krznaric, Roman, 'Why we need to reinvent democracy for the long-term', BBC Future (2020); Krznaric (2020).

45   Wellbeing of Future Generations Bill [HL] 2019–21.

46   John, T. M., and MacAskill, W., 'Longtermist Institutional Reform', in Cargill, Natalie and John, Tyler M. (ed.), *The Long View* (Longview Philanthropy, 2020).

47   Breckon, J., et al, 'Evidence vs Democracy: How "mini-publics" can traverse the gap between citizens, experts, and evidence', Alliance for Useful Evidence (2019).

48   Krznaric, Roman, 'Four ways to redesign democracy for future generations', *Open Democracy* (2020).

49   Saijō, T., 'Future Forebearers', *RSA Journal* (2021).

50   de Tocqueville (1838).

4. The Timekeeping Ape

1   Bergson, Henri, *L'Évolution créatrice* (1907), translated by Mitchell, Arthur, *Creative Evolution* (Henry Holt & Company, 1911).

2   In particular, the French philosopher Henri Bergson, who made a distinction between scientific, mathematical time and the human experience of time which he called 'real duration' (*durée réelle*).

3   Woolf, Virginia, *Mrs. Dalloway* (Harcourt, Brace & Co, 1925); Taunton, Matthew, 'Modernism, time and consciousness: the influence of Henri Bergson and Marcel Proust', *British Library: Discovering Literature* (2016).

4   Woolf, Virginia, *Orlando: A Biography* (Hogarth Press, 1928).

5   Osvath, M., 'Spontaneous planning for future stone throwing by a male chimpanzee', *Current Biology* (2009); Osvath, M., and Karvonen, E., 'Spontaneous Innovation for Future Deception in a Male Chimpanzee', *PLoS ONE* (2012).

6   Sample, Ian, 'Chimp who threw stones at zoo visitors showed human trait, says scientist', *Guardian* (2009).

7   Aristotle wrote that 'many animals have memory and are capable of instruction, but no other animal except man can recall the past at will'.

8   Nietzsche, Friedrich, *Untimely Meditations* (1873–1876; Cambridge University Press, 1997).

9   Mulcahy, N., 'Apes Save Tools for Future Use', *Science* (2006).

10   For a literature review of animal foresight, see Redshaw, J., and Bulley, A., 'Future-Thinking in Animals: Capacities and Limits', in Oettingen, Gabrielle, Timur, Sevincer and Gollwitzer, Peter (eds.), *The Psychology of Thinking about the Future* (Guilford Press, 2018).

11   Corballis, M. C., 'Mental time travel, language, and evolution', *Neuropsychologia* (2019).

12   Knolle, F., et al., 'Sheep Recognize Familiar and Unfamiliar Human Faces from Two-Dimensional Images', *Royal Society Open Science* (2017).

13   Roberts, W. A., 'Are animals stuck in time?', *Psychological Bulletin* (2002). Based on correspondence between Roberts and Michael D'Amato.

14   Russell, Bertrand, *Human Society in Ethics and Politics* (1954; Routledge, 2009).

15   Hublin, Jean-Jacques, et al., 'New fossils from Jebel Irhoud, Morocco and the pan-African origin of *Homo sapiens*', *Nature* (2017).

16   This is a big topic, so further reading might include: Henrich, Joseph, *Secrets of our Success: How Culture Is Driving Human Evolution, Domesticating Our Species, and Making Us Smarter* (Princeton University Press, 2015); Vince, Gaia, *Transcendence: How Humans Evolved through Fire, Language, Beauty, and Time* (Penguin, 2019).

17   'Q&A: Thomas Suddendorf', *Current Biology* (2015).

18   Suddendorf, Thomas, *Discovery of the Fourth Dimension: Mental Time Travel and Human Evolution* (Master's thesis, 1994).

19   While Suddendorf's proposal that mental time travel is uniquely human holds as a plausible theory, some scientists disagree over the details, citing the animal evidence covered earlier in the chapter. Suddendorf, to his credit, says he would embrace strong evidence of animal mental time travel if it emerged, not least because it would change our relationship with nature and our responsibilities towards animals.

20   Suddendorf provides a more detailed exploration of hominin mental time travel in his own book. See: Suddendorf, Thomas, *The Gap: The Science of What Separates Us from Other Animals* (Basic Books, 2013).

21 Tulving, E., 'Memory and consciousness', *Canadian Psychology* (1985); Terrace, Herbert S., and Metcalfe, Janet (eds.), *The Missing Link in Cognition: Origins of Self-Reflective Consciousness* (Oxford University Press USA, 2005); Rosenbaum, R., et al., 'The case of K. C.: contributions of a memory-impaired person to memory theory', *Neuropsychologia* (2005); interviews with K. C., available at: youtube.com/watch?v=tXHkoa3RvLc (accessed January 2020).

22 Suddendorf, T., and Busby, J., 'Making decisions with the future in mind: Developmental and comparative identification of mental time travel', *Learning and Motivation* (2005).

23 Tulving, E., 'Episodic Memory and Autonoesis: Uniquely Human?', in Terrace and Metcalfe (2005).

24 As told by Suddendorf in *The Gap* (p. 102).

25 Corballis, M., 'Language, Memory, and Mental Time Travel: An Evolutionary Perspective', *Frontiers in Human Neuroscience* (2019).

26 Seligman, M., et al., *Homo Prospectus* (Oxford University Press, 2016).

27 Kahneman, Daniel, *Thinking, Fast and Slow* (Penguin, 2012).

5. The Psychology of Yesterday, Today and Tomorrow

1 Hume, David, *An Enquiry Concerning the Principles of Morals* (A. Millar, 1751).

2 The original painting by Pieter Bruegel the Elder is lost, and the one now displayed is thought to be a copy by an unknown artist.

3 Forman-Barzilai, Fonna, *Adam Smith and the Circles of Sympathy* (Cambridge University Press, 2010).

4 Liberman, N., and Trope, Y., 'The Psychology of Transcending the Here and Now', *Science* (2008); Trope, Y., and Liberman, N., 'Construal-level theory of psychological distance', *Psychological Review* (2010).

5 Hanson, Robin, 'The Future Seems Shiny', *Overcoming Bias* (2010).

6 This list is compiled from Trope and Liberman's research, plus a collection of near and far effects described by Robin Hanson. See: Hanson, Robin, 'Near–Far Summary', *Overcoming Bias* (2010).

7 Hume, David, *A Treatise of Human Nature, Book III: 'Of Morals'* (John Noon, 1739).

8 Hershfield, H., 'Future self-continuity: how conceptions of the future self transform intertemporal choice', *Annals of the New York Academy of Sciences* (2011).

9   Pahl, S., and Bauer, J., 'Overcoming the Distance: Perspective Taking With Future Humans Improves Environmental Engagement', *Environment and Behavior* (2013).

10   Saijō, T., 'Future Forebearers', *RSA Journal* (2021).

11   'The Future Energy Lab', Superflux (2019).

12   Conant, Jennet, *109 East Palace: Robert Oppenheimer and the Secret City of Los Alamos* (Simon & Schuster, 2007); Achenbach, Joel, 'The man who feared, rationally, that he'd just destroyed the world', *Washington Post* (2015).

13   For more on expectation effects, see: Robson, David, *The Expectation Effect: How Your Mindset Can Transform Your Life* (Canongate, 2022).

14   Gerbner, G., 'The "Mainstreaming" of America: Violence Profile No. 11', *Journal of Communication* (1980). More recently, scientists have looked at the negative mental health outcomes, e.g. Pfefferbaum, B., et al., 'Disaster Media Coverage and Psychological Outcomes: Descriptive Findings in the Extant Research', *Current Psychiatry Reports* (2014).

15   Schelling, T., 'The Role of War Games and Exercises', in Carter, A., et al. (ed.) *Managing Nuclear Operations* (Brookings Institution, 1987).

16   Desvousges, W., et al., *Measuring Nonuse Damages Using Contingent Valuation: An Experimental Evaluation of Accuracy* (RTI Press, 2010).

17   'On caring', *Minding our way* (2014).

18   Fetherstonhaugh, D., et al., 'Insensitivity to the Value of Human Life: A Study of Psychophysical Numbing', *Journal of Risk and Uncertainty* (1997).

19   Jenni, K., and Loewenstein, G., 'Explaining the Identifiable Victim Effect', *Journal of Risk and Uncertainty* (1997).

20   First attributed to Stalin in the *Washington Post* (1947) as 'If only one man dies of hunger, that is a tragedy. If millions die, that's only statistics'; Mother Teresa quoted in Slovic, P., '"If I look at the mass I will never act": Psychic numbing and genocide', *Judgment and Decision Making* (2007).

21   Morton, Timothy, *Hyperobjects: Philosophy and Ecology after the End of the World* (University of Minnesota Press, 2013).

22   Markowitz, E., and Shariff, A., 'Climate change and moral judgement', *Nature Climate Change* (2012).

23   I owe a number of the details of Mainstone's story to Trent Dalton's reporting, and would recommend his full article: Dalton, Trent, 'Pitch Fever', *The Australian* in Hay, Ashley (ed.), *The Best Australian Science Writing 2014* (NewSouth Publishing, 2014).

24 'Humans Wired to Respond to Short-Term Problems', *Talk of the Nation*, NPR (2006).

25 Davies, T., 'Slow violence and toxic geographies: "Out of sight" to whom?', *Environment and Planning C: Politics and Space* (2021).

26 Nixon, Rob, *Slow Violence and the Environmentalism of the Poor* (Harvard University Press, 2013).

27 Svedäng, H., 'Long-term impact of different fishing methods on the ecosystem in the Kattegat and Öresund', Paper for European Parliament's Committee on Fisheries (2010).

28 Mowat, Farley, *Sea of Slaughter* (McClelland and Stewart, 1984).

29 Pauly, D., 'Anecdotes and the shifting baseline syndrome of fisheries', *Trends in Ecology and Evolution* (1995); Pauly, D., *Vanishing Fish: Shifting Baselines and the Future of Global Fisheries* (Greystone Books, 2019).

30 Kahn, P., 'Children's affiliations with nature: Structure, development, and the problem of environmental generational amnesia', in Kellert, Stephen, and Kahn, Peter (eds.), *Children and Nature* (MIT Press, 2002).

31 Soga, M., and Gaston, K., 'Shifting baseline syndrome: causes, consequences, and implications', *Frontiers in Ecology and the Environment* (2018); Jones, L., Turvey, S., Massimino, D., and Papworth, S., 'Investigating the implications of shifting baseline syndrome on conservation', *People and Nature* (2020); Moore, F., Obradovich, N., Lehner, F., and Baylis, P., 'Rapidly declining remarkability of temperature anomalies may obscure public perception of climate change', *Proceedings of the National Academy of Sciences* (2019).

32 Kahn, P., and Weiss, T., 'The Importance of Children Interacting with Big Nature', *Children, Youth and Environments* (2017).

33 Parker, Theodore, *Ten Sermons of Religion* (Crosby, Nichols & Co., 1853).

34 Tonn, B., Hemrick, A., and Conrad, F., 'Cognitive representations of the future: Survey results', *Futures* (2006). See also: 'The American Future Gap?', Institute for the Future (2017).

35 There were also slight differences between demographics surveyed: the religion of the person made a slight difference. People of the Jewish and traditional Asian faiths, for example, had longer future horizons than Christians or secular people.

36 Zhang, J. W., Howell, R. T., and Bowerman, T., 'Validating a brief measure of the Zimbardo Time Perspective Inventory', *Time and Society* (2013); original paper: Boyd, J., and Zimbardo, P., 'Putting time in perspective: A

valid, reliable individual-differences metric', *Journal of Personality and Social Psychology* (1999); book: Boyd, John, and Zimbardo, Philip, *The Time Paradox: The New Psychology of Time That Will Change Your Life* (Atria, 2009); a more recent review paper: Peng, C., et al., 'A Systematic Review Approach to Find Robust Items of the Zimbardo Time Perspective Inventory', *Frontiers in Psychology* (2021).

37   While psychologists have refined the original test, its results have been replicated in around 25 countries. See: Sircova A., et al., 'A global look at time: a 24-country study of the equivalence of the Zimbardo Time Perspective Inventory', *SAGE Open* (2014).

38   Strathman, A., et al., 'The consideration of future consequences: Weighing immediate and distant outcomes of behaviour', *Journal of Personality and Social Psychology* (1994); Husman, J., and Shell, D. F., 'Beliefs and perceptions about the future: A measurement of future time perspective', *Learning and Individual Differences* (2008).

39   Milfont, T., Wilson, J., and Diniz, P., 'Time perspective and environmental engagement: A meta-analysis', *International Journal of Psychology* (2012).

40   Carelli, M. G., Wiberg, B., and Wiberg, M., 'Development and construct validation of the Swedish Zimbardo Time Perspective Inventory', *European Journal of Psychological Assessment* (2011).

41   Rönnlund, M., et al., 'Mindfulness Promotes a More Balanced Time Perspective: Correlational and Intervention-Based Evidence', *Mindfulness* (2019).

42   Boniwell, I., Osin, E. N., and Sircova, A., 'Introducing time perspective coaching: A new approach to improve time management and enhance well-being', *International Journal of Evidence Based Coaching* (2014).

43   Lamm, B., et al., 'Waiting for the Second Treat: Developing Culture-Specific Modes of Self-Regulation', *Child Development* (2017).

44   Benjamin, D., et al., 'Predicting mid-life capital formation with pre-school delay of gratification and life-course measures of self-regulation', *Journal of Economic Behavior and Organization* (2020).

45   The journalist Bina Venkataraman discusses the implications of these marshmallow experiments in more detail. See: Venkataraman, Bina, *The Optimist's Telescope: Thinking Ahead in a Reckless Age* (Riverhead Books, 2019).

46   'National Culture', *Hofstede Insights*; Hofstede, Geert, Hofsted, Gert Jan, and Minkov, Michael, *Cultures and Organizations: Software of the Mind* (Macgraw-Hill Education, Third Edition, 2010).

47 According to Hofstede's scores: US (26), UK (51), Australia (21), Japan (88), China (87), Russia (81). Source: Hofstede Insights, Country Comparison, hofstede-insights.com/country-comparison.

48 Galor, O., Özak, Ö., and Sarid, A., 'Geographical origins and economic consequences of language structures', *CESifo Working Paper Series No. 6149* (2016).

49 e.g. Grabb, E., Baer. D., and Curtis, J., 'The Origins of American Individualism: Reconsidering the Historical Evidence', *The Canadian Journal of Sociology* (1999).

50 Doebel, S., and Munakata, Y., 'Group Influences on Engaging Self-Control: Children Delay Gratification and Value It More When Their In-Group Delays and Their Out-Group Doesn't', *Psychological Science* (2018).

51 Pryor, C., Perfors, A., and Howe, P., 'Even arbitrary norms influence moral decision-making', *Nature Human Behaviour* (2018).

52 However, it works both ways.

53 Burger, J., et al., 'Nutritious or delicious? The effect of descriptive norm information on food choice', *Journal of Social and Clinical Psychology* (2010); Wenzel, M., 'Misperceptions of social norms about tax compliance: From theory to intervention', *Journal of Economic Psychology* (2005); 'Applying Behavioural Insights to Organ Donation: preliminary results from a randomised controlled trial', UK Cabinet Office (2013).

54 Wade-Benzoni, K. A., 'A golden rule over time: Reciprocity in intergenerational allocation decisions', *Academy of Management Journal* (2002); Bang, H. M., et al., 'It's the thought that counts over time: The interplay of intent, outcome, stewardship, and legacy motivations in intergenerational reciprocity', *Journal of Experimental Social Psychology* (2017).

55 Watkins, H., and Goodwin, G., 'Reflecting on Sacrifices Made by Past Generations Increases a Sense of Obligation Towards Future Generations', *Personality and Social Psychology Bulletin* (2020). A caveat to note is that, in this study, a sense of moral obligation was increased by making these past reflections, but not necessarily the willingness to make actual financial sacrifices.

56 Zaval, L., et al., 'How Will I Be Remembered? Conserving the Environment for the Sake of One's Legacy', *Psychological Science* (2015).

57 Bain, P. G., et al., 'Collective Futures: How Projections About the Future of Society Are Related to Actions and Attitudes Supporting Social Change', *Personality and Social Psychology Bulletin* (2013).

6. Long-terminology

1   Núñez, R., et al., 'Contours of time: Topographic construals of past, present, and future in the Yupno valley of Papua New Guinea', *Cognition* (2012).

2   Cooperrider, Kensy, and Núñez, Rafael, 'How We Make Sense of Time', *Scientific American* (2016).

3   Cooperrider, K., Slotta, J., and Núñez, R., 'Uphill and Downhill in a Flat World: The Conceptual Topography of the Yupno House', *Cognitive Science* (2016).

4   Kant, Immanuel, *Anthropology from a Pragmatic Point of View* (1798, Cambridge University Press, 2006).

5   Dor, Daniel, *The Instruction of Imagination: Language as a Social Communication Technology* (Oxford University Press, 2015).

6   Fuhrman, O., et al., 'How Linguistic and Cultural Forces Shape Conceptions of Time: English and Mandarin Time in 3D', *Cognitive Science* (2011).

7   As do speakers of Guugu Yimithirr, another Aborigine language.

8   Boroditsky, L., and Gaby, A., 'Remembrances of Times East', *Psychological Science* (2010).

9   Boroditsky, L., 'How Languages Construct Time', in Dehaene, Stanislas, and Brannon, Elizabeth (eds.), *Space, Time and Number in the Brain: Searching for the Foundations of Mathematical Thought* (Elsevier, 2011).

10   Núñez, R., and Sweetser, E., 'With the Future Behind Them: Convergent Evidence From Aymara Language and Gesture in the Crosslinguistic Comparison of Spatial Construals of Time', *Cognitive Science* (2006).

11   Conceptually, it's also not dissimilar to the Maori idea of walking backward into the future.

12   Dahl, O., 'When the future comes from behind: Malagasy and other time concepts and some consequences for communication', *International Journal of Intercultural Relations* (1995).

13   Looking over the 'left shoulder' to see the immediate future is also described in Aymara.

14   Radden, G., 'The Metaphor TIME AS SPACE across Languages', *Zeitschrift Für Interkulturellen Fremdsprachenunterricht* (2015).

15   Via correspondence with Phillippe Lemonnier at Pacific Ventury, Tahitian speaker.

16   Fuhrman (2011).

17   Sinha, C., et al., 'When Time Is Not Space: The Social and Linguistic Construction of Time Intervals and Temporal Event Relations in an Amazonian Culture', *Language and Cognition* (2014).

18   Whorf, B. L., 'An American Indian Model of the Universe', *International Journal of American Linguistics* (1950).

19   Malotki, Ekkehart, *Hopi Time: A Linguistic Analysis of the Temporal Concepts in the Hopi Language* (Mouton de Gruyter, 1983).

20   For a longer list of untranslatable words that relate to emotion, see psychologist Tim Lomas's 'positive lexography' project. Available at: www.drtimlomas.com/lexicography.

21   Leane, Jeanine, *Guwayu – For All Times: A Collection of First Nations Poems* (Magabala Books, 2020).

22   Deutscher, Guy, *Through the Language Glass: Why the World Looks Different in Other Languages* (Metropolitan Books/Henry Holt & Company, 2010).

23   Haviland, J., 'Anchoring, Iconicity, and Orientation in Guugu Yimithirr Pointing Gestures', *Journal of Linguistic Anthropology* (1993).

24   de Silva, Mark, 'Guy Deutscher on *"Through the Language Glass"*', *The Paris Review* (2010).

25   Boroditsky, L., Schmidt, L., and Phillips, W., 'Sex, syntax, and semantics', in Gentner, Dedre, and Goldin-Meadow, Susan (eds.), *Language in Mind: Advances in the Study of Language and Thought* (MIT Press, 2003).

26   Similarly for the word 'key', which is masculine in German and feminine in Spanish. Germans described keys as hard, heavy, jagged, metal, serrated and useful. Spanish speakers used golden, intricate, little, lovely, shiny and tiny. However, it should be noted that other researchers have tried to replicate these findings and been unsuccessful. See: Mickan, A., Schiefke, M., and Anatol, S., 'Key is a llave is a Schlüssel: A failure to replicate an experiment from Boroditsky et al.', in Hilpert, M., and Flach, S. (eds.), *Yearbook of the German Cognitive Linguistics Association* (Walter de Gruyter, 2003).

27   A longer list: *Strong future*: English, French, Italian, Spanish, Portuguese, Turkish, Arabic, Hebrew, Russian, Bengali, Gujarati, Hindi, Kashmiri, Panjabi, Urdu, most Eastern European languages, Korean, Thai. *Weak future*: German, Danish, Dutch, Flemish, Icelandic, Norwegian, Swedish, Estonian, Indonesian, Japanese, Malay, Maori, Sudanese, Vietnamese, Cantonese, Mandarin.

28   Chen, M., 'The Effect of Language on Economic Behavior: Evidence from Savings Rates, Health Behaviors, and Retirement Assets', *American Economic Review* (2013).

29   Beckwith, S., and Reed, J., 'Impounding the Future: Some Uses of the Present Tense in Dickens and Collins', *Dickens Studies Annual* (2002).

30   Roberts, S. G., Winters, J., and Chen, K., 'Future Tense and Economic Decisions: Controlling for Cultural Evolution', *PLoS ONE* (2015).

31   Chen, S., et al., 'Languages and corporate savings behavior', *Journal of Corporate Finance* (2017); Liang, H., et al., 'Future-time framing: The effect of language on corporate future orientation', *Organization Science* (2018).

32   Mavisakalyan, A., Tarverdi, Y., and Weber, C., 'Talking in the present, caring for the future: Language and environment', *Journal of Comparative Economics* (2018); Kim, S., and Filimonau, V., 'On linguistic relativity and pro-environmental attitudes in tourism', *Tourism Management* (2017); Pérez, E. O., and Tavits, M., 'Language shapes people's time perspective and support for future-oriented policies: Language and political attitudes', *American Journal of Political Science* (2017).

33   Sutter, M., Angerer, S., Glätzle-Rützler, D., and Lergetporer, P., 'Language group differences in time preferences: Evidence from primary school children in a bilingual city', *European Economic Review* (2018).

34   Ayres, I., Kricheli Katz, T., and Regev, T., 'Do Languages Generate Future-Oriented Economic Behavior?', SSRN (2020).

35   Stanner, W. E. H., *The Dreaming and Other Essays* (Black Inc. Agenda, 2011).

36   Thibodeau, P., and Boroditsky, L., 'Metaphors We Think With: The Role of Metaphor in Reasoning', *PLoS ONE* (2011).

37   Ewieda, S., 'The realization of time metaphors and the cultural implications: An analysis of the Quran and English Quranic translations' (unpublished thesis, 2006).

38   Based on a simple search in Google Ngram: while 'time to kill' or the idea of 'beating time' can be found in books in the 1800s, time as 'a bitch' or 'enemy' only seemed to have emerged in the twentieth century.

39   Boroditsky, 'How Languages Construct Time' (2011).

7. A Delightful Horror

1   Burke, Edmund, *A Philosophical Enquiry into the Origins of the Sublime and Beautiful: And Other Pre-Revolutionary Writings* (1757; Penguin Classics, 1998).

2   von Baer, K. E., *Welche Auffassung der lebenden Natur ist die richtige?* (1862).

I owe the discovery of von Baer's thought experiment to: Burdick, Alan, *Why Time Flies: A Mostly Scientific Investigation* (Simon & Schuster, 2017).

3   Lyell, Charles, *Principles of Geology* (1830–33; Penguin Classics, 1998).

4   Close readers may note Lyell's use of the word 'infinite' echoes the faith-based perspective of eternal time. Hutton too framed time as being 'without end'. So, while these early geologists made discoveries that unlocked deep time, they themselves would seem to have had a long view that was entwined with the dominant religious perspective.

5   McPhee, John, *Basin and Range* (Farrar, Straus and Giroux, 1981).

6   Burke, Edmund, *Reflections on the Revolution in France* (J. Dodsley, 1790).

7   Obviously the more isolated a population (e.g. a remote tribe), the further back you have to go, but given the amount of migration over the centuries, this applies to the majority of people on Earth.

8   'Historical Estimates of World Population', United States Census (2021).

9   Rutherford, Adam, *A Brief History of Everyone Who Ever Lived: The Stories in Our Genes* (Weidenfeld & Nicolson, 2016).

10   Rohde, D., Olson, S., and Chang, J., 'Modelling the recent common ancestry of all living humans', *Nature* (2004); Ralph, P., and Coop, G., 'The Geography of Recent Genetic Ancestry Across Europe', *PLoS Biology* (2013); Hein, J., 'Pedigrees for all humanity', *Nature* (2004).

11   MacAskill, W., and Mogensen, A., 'The paralysis argument', *Philosophers' Imprint*, Global Priorities Institute Working Paper (2019).

12   Aschenbrenner, Leopold, 'Burkean Longtermism', *For Our Posterity* (2021).

13   A caveat: Burke's focus on posterity here was really reaching forward to living children. So, strictly, the most precise term might be 'Burkean-inspired'.

14   Burke (1757; 1998); Frank, J., '"Delightful Horror": Edmund Burke and the Aesthetics of Democratic Revolution' (unpublished paper, 2014).

15   Carlyon, Clement, *Early Years and Late Reflections* (Routledge, 1936).

16   Kant, Immanuel, *Kritik der Urteilskraf* (1790), translated by Guyer, Paul, and Matthews, Eric, *Critique of the Power of Judgment* (Cambridge University Press, 2002).

17   Wordsworth, W., 'Lines Written a Few Miles above Tintern Abbey' in *Lyrical Ballads With a Few Other Poems* (J. & A. Arch, 1798).

18   Bjornerud, Marcia, *Timefulness: How Thinking Like a Geologist Can Help Save the World* (Princeton University Press, 2018).

19   Macfarlane, Robert, *Underland: A Deep Time Journey* (Penguin, 2019).

20   von Baer (1862), translated by Carlsberg, Karl, 'A Microscope for Time:

What Benjamin and Klages, Einstein and the Movies Owe to Distant Stars',
in Miller, Tyrus (ed.), *Given World and Time: Temporalities in Context* (Central
European University Press, 2008).

8. Timeviews

1   Thompson, E.P., 'Time, Work-Discipline, and Industrial Capitalism', *Past and Present* (1967).

2   Franklin, B., 'Advice to a Young Tradesman' in Fisher, George, *The American Instructor, Or, Young Man's Best Companion* (B. Franklin and D. Hall, 1748).

3   'The Zarathusti World: A 2012 demographic picture', *The Federation of Zoroastrian Associations of North America (FEZANA)* (2012).

4   Stewart, Sarah, Hintze, Almut, and Williams, Alan (eds.), *The Zoroastrian Flame: Exploring Religion, History and Tradition* (Bloomsbury Publishing, 2016).

5   Hornsby, David, 'The Zoroastrian Flame', *Beshara Magazine* (2018).

6   Two early eighth-century documents, the Kojiki and Nihonshoki, tell stories of specific kami. And a tenth-century text Engishiki describes rituals.

7   'Rebuilding Every 20 Years Renders Sanctuaries Eternal – the Sengu Ceremony at Jingu Shrine in Ise', JFS Japan for Sustainability (2013).

8   Smith, Daigo, 'Traditions: Shikinen Sengu', Japan Woodcraft Association (2020).

9   Adams, C., 'Japan's Ise Shrine and Its Thirteen-Hundred-Year-Old Reconstruction Tradition', *Journal of Architectural Education* (1998).

10 Rose, Alexander, 'Long-term Building in Japan', The Long Now Foundation (2019).

11 For long-mindedness, the Zoroastrians might not get a perfect report card. In India, men in mixed marriages can bring their children into the faith, but it's not universally accepted by the most ardent traditionalists. And if women marry out, then it's game over for the line. To prevent the religion slowly shrinking in the long term, its leaders may eventually need to relax these membership rules. Still, the faith has navigated worse trials over the centuries.

12 *Extraordinary Rituals – Why Would You Do This?*, BBC (2018).

13 Whitehouse, H., and Lanman, J. A., 'The Ties That Bind Us: Ritual, Fusion, and Identification', *Current Anthropology* (2014).

14 The last Lord Mallard, in 2001, was Martin Litchfield West, who died in 2015. By coincidence, one of his areas of scholarship was Zoroastrianism.

15 'The Mallard Society', All Souls College (2001).

16 Chai, D., 'Zhuangzi's Meontological Notion of Time', *Dao* (2014); Jhou, N., 'Daoist Conception of Time: Is Time Merely a Mental Construction?', *Dao* (2020).

17 Kalupahana, D., 'The Buddhist Conception of Time and Temporality', *Philosophy East and West* (1974); 'What are kalpas?', *Lion's Roar* (2016); Maguire, Jack, *Essential Buddhism* (Atria, 2001).

18 Ijjas, Anna, 'What if there was no big bang and we live in an ever-cycling universe?', *New Scientist* (2019).

19 Thompson, E. P. (1967).

20 Janca, A., and Bullen, C., 'The Aboriginal concept of time and its mental health implications', *Australasian Psychiatry* (2003).

21 'Dibang Valley case study', *Flourishing Diversity* (2021).

22 Robbins, Jim, 'Native Knowledge: What Ecologists Are Learning from Indigenous People', *Yale E360* (2018).

23 Huntingdon, H., and Mymrin, N., 'Traditional Knowledge of the Ecology of Beluga Whales', *Arctic* (1999).

24 Author unknown, *The Constitution of the Iroquois Nations* (Kessinger Publishing, 2010).

25 Deloria Jr, Vine, 'American Indians and the Moral Community', in Deloria Jr, Vine, and Treat, James, *For This Land: Writings on Religion in America* (Routledge, 1998).

26 Wilkins, David E., 'How to Honor the Seven Generations', *Indian Country Today* (2015).

9. Longtermism

1   Sometimes attributed to Groucho Marx, this quip probably predates him. As Joseph Addison in *The Spectator* in 1714 wrote:
    'We are always doing, says he, something for posterity, but I would fain see posterity do something for us.'

2   Ramsey, F., 'A Mathematical Theory of Saving', *The Economic Journal* (1928).

3   Beard, S. J., 'Parfit Bio' (2020), sjbeard.weebly.com; 'S. J. Beard on Parfit, Climate Change, and Existential Risk', *Hear This Idea* (2020).

4   Taebi, B., and Kloosterman, J., 'To Recycle or Not to Recycle? An Intergenerational Approach to Nuclear Fuel Cycles', *Science and Engineering Ethics* (2007).

5   Parfit, Derek, *Reasons and Persons* (Oxford University Press, 1984).

6   Parfit, Derek, *On What Matters: Volume II* (Oxford University Press, 2011).

7   Parfit, Derek, *On What Matters: Volume III* (Oxford University Press, 2017).

8   'Toby Ord: Why I'm giving £1m to charity', BBC News (2010).

9   For example, one criticism is that it only directs money to what can be *measured*. Some feel that neglects causes that cannot be turned into comparative data-tables and rankings.

10 Beckstead, N., 'On the Overwhelming Importance of Shaping the Far Future' (unpublished doctorate dissertation, 2013).

11 Schubert, S., Caviola, L., and Faber, N., 'The Psychology of Existential Risk: Moral Judgments about Human Extinction', *Scientific Reports* (2019).

12 Parfit (1984).

13 Ord, Toby, *The Precipice: Existential Risk and the Future of Humanity* (Bloomsbury, 2020).

14 'The Green Book: Central Government Guidance on Appraisal and Evaluation', HM Treasury (2020); 'Intergenerational wealth transfers and social discounting: Supplementary Green Book guidance', HM Treasury (2008).

15 Ramsey, F., 'A Mathematical Theory of Saving', *The Economic Journal*, (1928).

16 For example, see the 'Nordhaus vs Stern' debate: Nordhaus, W., 'A Review of the Stern Review on the Economics of Climate Change', *Journal of Economic Literature* (2007); Stern, N., *The Economics of Climate Change: The Stern Review* (LSE, 2006).

17 Cowen, T., and Parfit, D., 'Against the social discount rate', in Laslett, Peter, and Fishkin, James (eds.), *Justice Between Age Groups and Generations* (New Haven, 1992).

18 *Future Proof: The opportunity to transform the UK's resilience to extreme risks*, The Centre for Long-term Resilience (2021).

19 A note on my assumptions: These numbers were calculated using the average global fertility rate projected for this century – approximately two children per woman. I chose 50,000 years because that is roughly the same period of time that we are from the first anatomically modern humans with language. But all calculations are based on a few broad assumptions, because projecting population into the far future is always going to be speculative. In the West, population may be on the cusp of

slow decline, while in the developing world it is projected to grow significantly this century, especially in countries like Nigeria, which by 2100 may have as big a population as Europe and North America combined. (See: Rees, M., 'Some Thoughts on 2050 and Beyond', *American Philosophical Society* (2021).) Researchers at the United Nations once produced a report attempting to model population all the way up to 2300. In their 'medium' scenario, they concluded that 'world population growth beyond 2050, at least for the following 250 years, is expected to be minimal'. (See: *World Population to 2300*, United Nations (2004).) Therefore, for the sake of illustration – and to satisfy my own curiosity – I made the assumption that, over the very long term, the average number of new people born per century will stabilise. It may well rise, in which case the numbers would be even bigger. But even if it begins to fall off, global population will still remain very large for a long time, unless we were plummeting towards extinction.

20 Newberry, T., 'How many lives does the future hold?', *Global Priorities Institute Technical Report T2-2021* (2021).

21 Bostrom, Nick, *Superintelligence: Paths, Dangers, Strategies* (Oxford University Press, 2014).

22 Cremer, Z. C., and Kemp, L., 'Democratising Risk: In Search of a Methodology to Study Existential Risk', *Arxiv* (2021).

23 Assuming paper stock of 8gsm, which is about 100 microns thick.

24 Assuming a Bible is 1,200 pages, *Great Expectations* ~550 pages and *Communist Manifesto* ~44 pages.

25 Via Sbiis Saibian's 'Large Number Site'.

26 'The Green Pea Analogy', Maxstudy.org.

27 MacAskill, William, *What We Owe the Future: A Million-Year View* (Oneworld, 2022).

28 Snyder-Beattie, A. E., Ord, T., and Bonsall, M. B., 'An upper bound for the background rate of human extinction', *Scientific Reports* (2019).

29 Torres, E. P., 'Against longtermism', *Aeon* (2021).

30 Jan Narveson: 'We are in favor of making people happy, but neutral about making happy people.' See Narveson, J., 'Moral problems of population', *The Monist* (1973).

31 Masrani, Vaden, 'A Case Against Strong Longtermism', vmasrani.github.io (2021).

32 Holt, Jim, 'The Power of Catastrophic Thinking', *The New York Review of Books* (2021).

33   Singer, P., 'The Hinge of History', *Project Syndicate* (2021).

34   Thorstad, D., 'The scope of longtermism', GPI Working Paper No. 6-2021 (2021).

35   MacAskill (2022).

36   Ord was commenting in a discussion on the Effective Altruism Forum. See: 'Towards a weaker longtermism', *EA Forum* (2021).

10. Temporal Windows

1   Wittgenstein, Ludwig, *Philosophische Untersuchungen* (1953), translated by Anscombe, Gertrude E. M., *Philosophical Investigations* (Macmillan, 1958).

2   Burroughs, John, *The Complete Nature Writings of John Burroughs* (William H. Wise & Company, 1913).

3   Stearns, Stephen C., 'Lecture 1. The Nature of Evolution: Selection, Inheritance, and History', *Open Yale Courses: Principles of Evolution, Ecology and Behavior* (2009).

4   'The Elements of Life Mapped Across the Milky Way' by SDSS/APOGEE', SDSS (2017).

5   Tolkien, J. R. R., *On Fairy-Stories* (Oxford University Press, 1947); Cotton-Barratt, O., and Ord, T., 'Existential Risk and Existential Hope: Definitions', *Future of Humanity Institute – Technical Report* (2015).

6   Sagan, Carl, *Broca's Brain: Reflections on the Romance of Science* (Random House, 1979).

7   'Pigeon waste, cosmic melodies and noise in scientific communication', *Lindau Nobel Laureate Meetings* (2010).

8   Faisal ur Rahman, Syed, 'The enduring enigma of the cosmic cold spot', *Physics World* (2020); An, D., et al., 'Apparent evidence for Hawking points in the CMB Sky', *Monthly Notices of the Royal Astronomical Society* (2020).

9   Davies, H., et al., 'Back to the future: Testing different scenarios for the next supercontinent gathering', *Global and Planetary Change* (2018).

10   Davies (2018).

11   Stafford, Tom, 'Reasons to trust models', *Reasonable People* (2020).

12   Andermann, T., et al., 'The past and future human impact on mammalian biodiversity', *Science Advances* (2020).

13   Roberts, N., et al., 'Europe's lost forests: a pollen-based synthesis for the last 11,000 years', *Scientific Reports* (2018).

14 Vavrus, S., et al., 'Glacial Inception in Marine Isotope Stage 19: An Orbital Analog for a Natural Holocene Climate', *Scientific Reports* (2018).

15 Merheb, M., et al., 'Mitochondrial DNA, a Powerful Tool to Decipher Ancient Human Civilization from Domestication to Music, and to Uncover Historical Murder Cases', *Cells* (2019); Bennett, C., et al., 'The broiler chicken as a signal of a human reconfigured biosphere', *Royal Society Open Science* (2018).

16 Irving-Pease, E., et al., 'Rabbits and the Specious Origins of Domestication', *Trends in Ecology and Evolution* (2018).

17 Tait, C., et al., 'Sensory specificity and speciation: a potential neuronal pathway for host fruit odour discrimination in Rhagoletis pomonella', *Proceedings of the Royal Society B: Biological Sciences* (2016).

18 Kettlewell, H., 'Evolution of melanism: The study of a recurring necessity', *Clarendon* (1973); Antonovics, J., et al., 'Evolution in closely adjacent plant populations VIII. Clinal patterns at a mine boundary', *Heredity* (1970).

19 The Anthropocene Working Group voted for this marker in 2018. The AWG is a component body of the Subcommission on Quaternary Stratigraphy which is itself is part of the International Commission on Stratigraphy.

20 As voted by the Anthropocene Working Group in 2016, part of the International Commission on Stratigraphy. See Steffen, W., et al., 'The trajectory of the Anthropocene: The Great Acceleration', *The Anthropocene Review* (2015).

21 Elhacham, E., et al., 'Global human-made mass exceeds all living biomass', *Nature* (2020).

22 'Living Planet Report', World Wildlife Fund (2018).

23 Cardenas, L., et al., 'First mussel settlement observed in Antarctica reveals the potential for future invasions', *Scientific Reports* (2020); Lundgren, E., et al., 'Introduced herbivores restore Late Pleistocene ecological functions', *Proceedings of the National Academy of Sciences* (2020).

24 Campbell-Staton, S., et al., 'Ivory poaching and the rapid evolution of tusklessness in African elephants', *Science* (2021); Pigeon, G., et al., 'Intense selective hunting leads to artificial evolution in horn size', *Evolutionary Applications* (2016); Sanderson, S., et al., 'The pace of modern life, revisited', *Molecular Ecology* (2021).

25 Hazen, R., et al., 'On the mineralogy of the "Anthropocene Epoch"', *American Mineralogist* (2017); Corcoran, P., Moore C., and Jazvac, K., 'An anthropogenic marker horizon in the future rock record', *GSA Today* (2014).

26   Kemp, Luke, 'Are we on the road to civilisation collapse?', BBC Future (2019).

27   Steffen, W., et al., 'The trajectory of the Anthropocene: The Great Acceleration', *The Anthropocene Review* (2015).

28   Jackson, R., 'Eunice Foote, John Tyndall and a question of priority', *Notes and Records* (2019).

29   Tyndall, J., 'On the Transmission of Heat of different qualities through Gases of different kinds', *Notices of the Proceedings at the meetings of the members of the Royal Institution* (1859).

30   Aarhenius, S., 'On the Influence of Carbonic Acid in the Air upon the Temperature of the Ground', *Philosophical Magazine and Journal of Science* (1896).

31   Chamberlin, T. C., 'An Attempt to Frame a Working Hypothesis of the Cause of Glacial Periods on an Atmospheric Basis', *The Journal of Geology* (1899).

32   Weart, Spencer, *The Discovery of Global Warming* (Harvard University Press, 2008).

33   Callendar, G., 'The artificial production of carbon dioxide and its influence on temperature', *Quarterly Journal of the Royal Meteorological Society* (1938).

34   Plass, G., 'The Carbon Dioxide Theory of Climatic Change', *Tellus* (1956); 'Carbon Dioxide and the Climate', *American Scientist* (1956); 'Does Science Progress? Gilbert Plass Redux', *American Scientist* (2010).

35   Weart (2008).

36   Wittgenstein (1953; 1958); Sergio Sismondo, *An Introduction to Science and Technology Studies* (Wiley, 2011).

37   'The Peoples' Climate Vote', UNDP (2021).

38   Mugny, G., and Pérez, J., 'L'influence sociale comme processus de changement', *Hermes, La Revue* (1989); Fisher, R., '"Social cryptomnesia": How societies steal ideas', BBC Future (2020).

39   Næss, A., 'The Shallow and the Deep, Long-Range Ecology Movement: A Summary', *Inquiry* (1973).

11. The Persuasive Power of Symbols and Stories

1   Bergson, Henri, *Le rire: Essai sur la signification du comique* (Revue de Paris, 1900) translated by Brereton, Cloudesley, and Rothwell, Fred, *Laughter: An Essay on the Meaning of the Comic* (Macmillan, 1914).

2   Cameron, David, 'Leader's Speech', Conservative Party Conference, *British Political Speech* (2013).

3   Schama, Simon, 'The tree that shaped Britain', BBC News Magazine (2010).

4   'DIO's commitment to planting trees', *Inside DIO* (2020).

5   Fox, James, 'The mystery of the "free-range sculpture" that simply disappeared', *Christie's Magazine* (2016).

6   McMullan, M., et al., 'The ash dieback invasion of Europe was founded by two genetically divergent individuals', *Nature Ecology and Evolution* (2018).

7   Boyd, I., et al., 'The Consequence of Tree Pests and Diseases for Ecosystem Services', *Science* (2013).

8   Atkinson, Nick, 'Ash dieback: one of the worst tree disease epidemics could kill 95% of UK's ash trees', *The Conversation* (2019).

9   Morris, Steven, '"I hope people will find it joyful": David Nash exhibition opens in Cardiff', *Guardian* (2019).

10   Whitman, Walt, *Specimen Days and Collect* (Rees Welsh & Co, 1882).

11   The Long Time Project's logo is based on tree rings. Roman Krznaric writes of 'acorn brain' as a description of long-term thinking. The Long Now Foundation's unofficial mascot is the bristlecone pine.

12   Wilson, Matthew, 'Butterflies: The ultimate icon of our fragility', BBC Culture (2021).

13   Bateson (1994).

14   *Building the Ministry for the Future*, Chelsea Green Publishing/School of International Futures (2021).

15   McPhee, John, *Basin and Range* (Farrar, Straus and Giroux, 1981).

16   Pratt, J. M., 'A Fossil Guide to Mark Twain's Essay "Was the World Made for Man?"', *The Mark Twain Annual* (2005).

17   Orwell, George, 'Mark Twain – The Licensed Jester', *Tribune* (1943).

18   Thanks to Thomas Moynihan; Jeans, James Hopwood, *The Universe Around Us* (Cambridge University Press, 1929).

19   Auger, James, 'Superflux: Tools and methods for making change', *Speculative Edu* (2019).

20   Gray, Richard, 'The world's knowledge is being buried in a salt mine', BBC Future (2016).

21   Starks, Gavin, 'Longplayer: How Long Will We Be Long?', *Taking Time* (2021).

22   Ecker, Bogomir, *Die Tropfsteinmaschine, 1996–2496* (Hatje Cantz, 1996).

23   '"Fairytale of New York" the most played Christmas track of the 21st Century', *PPL* (2020).

24　　Russell, Andrew, and Vinsel, Lee, 'Hail the maintainers', *Aeon* (2016).

25　　'Defining Core Values', The Maintainers (2019).

26　　Eno, Brian, 'The Big Here and the Long Now', The Long Now Foundation (2009).

27　　The founding board also included Douglas Carlston, Esther Dyson, Kevin Kelly, Paul Saffo and Peter Schwartz.

28　　Rose, Alexander, 'How to build something that lasts 10,000 years', BBC Future (2019).

29　　Liu, G., et al., 'Factors influencing the service lifespan of buildings: An improved hedonic model', *Habitat International* (2014)

30　　Fox (2016).

12. A Deep Civilisation

1　　Stuart Mill, John, 'The Malt Duty', Hansard HC Deb, Volume 182 (17 April 1866).

2　　Karnofsky, Holden, 'Has Life Gotten Better?', *Cold Takes* (2021).

3　　Salk, J., 'Are We Being Good Ancestors?', *World Affairs: The Journal of International Issues* (1992).

4　　Ord, Toby, *The Precipice: Existential Risk and the Future of Humanity* (Bloomsbury, 2020).

5　　Quoidbach J., Gilbert D., and Wilson T., 'The end of history illusion', *Science* (2013).

6　　The 'end of history' was a term coined in the 1990s by the historian Francis Fukuyama to describe a form of political stasis, specifically how Western liberal democracy would become the final form of government and 'the end-point of mankind's ideological evolution'. This idea has since been challenged.

Index

Page references for notes are followed by n